SpringerBriefs in Energy

SpringerBriefs in Energy presents concise summaries of cutting-edge research and practical applications in all aspects of Energy. Featuring compact volumes of 50 to 125 pages, the series covers a range of content from professional to academic. Typical topics might include:

- A snapshot of a hot or emerging topic
- A contextual literature review
- A timely report of state-of-the art analytical techniques
- An in-depth case study
- A presentation of core concepts that students must understand in order to make independent contributions.

Briefs allow authors to present their ideas and readers to absorb them with minimal time investment.

Briefs will be published as part of Springer's eBook collection, with millions of users worldwide. In addition, Briefs will be available for individual print and electronic purchase. Briefs are characterized by fast, global electronic dissemination, standard publishing contracts, easy-to-use manuscript preparation and formatting guidelines, and expedited production schedules. We aim for publication 8–12 weeks after acceptance.

Both solicited and unsolicited manuscripts are considered for publication in this series. Briefs can also arise from the scale up of a planned chapter. Instead of simply contributing to an edited volume, the author gets an authored book with the space necessary to provide more data, fundamentals and background on the subject, methodology, future outlook, etc.

SpringerBriefs in Energy contains a distinct subseries focusing on Energy Analysis and edited by Charles Hall, State University of New York. Books for this subseries will emphasize quantitative accounting of energy use and availability, including the potential and limitations of new technologies in terms of energy returned on energy invested.

Tareq Mahbub

Encouraging Foreign Direct Investment (FDI) in Bangladesh's Power Sector

The key factors for long-term investment sustainability

 Springer

Tareq Mahbub
Faculty, Graduate Business Department
Kwantlen Polytechnic University
Surrey, BC, Canada

ISSN 2191-5520   ISSN 2191-5539   (electronic)
SpringerBriefs in Energy
ISBN 978-3-031-27989-8   ISBN 978-3-031-27990-4   (eBook)
https://doi.org/10.1007/978-3-031-27990-4

This Springer imprint is published by the registered company Springer Nature Switzerland AG
The registered company address is: Gewerbestrasse 11, 6330 Cham, Switzerland

Preface

Dear Readers:

This book is based on an empirical study conducted by the author in Bangladesh's power sector. This study has taken into consideration a wide range of institutional variables invoking the locational determinants of FDI, thus widening the prospects from four broad categories of investment dimensions, namely regulatory, economic and financial, political, and societal, for conducting FDI. In this way it has contributed to the extant literature on FDI in the power sector of developing countries by moving beyond the regulatory mechanism of rules and regulations and investment incentives to draw FDI. It could be argued that having a wider range of factors will help foreign investors to take on a more holistic view on considering FDI into the power sector of developing countries adding to their repertoire the key factors that influence investment decision-making. Additionally, this book informs governments to pick up the right signals to remove the obstacles for FDI into the power sector, thus producing an enabling environment for guiding additional FDI into their respective economies.

This book contains six chapters. Chapter one presents the background that sets the theme of this study especially depicting Bangladesh's "Vision 2021" to become a middle-income country by 2021 and have universal power for all. It depicts how power is shown as a principal catalyst for increasing Bangladesh's GDP growth set on a path of rapid industrialization and export oriented dynamic economic growth. Increased foreign direct investment (FDI) in the Bangladeshi power sector is seen as a critical pathway for a significant increase in power generation capacity to meet its future developmental growth targets. Chapter two is entitled "Benefits and Costs of FDI." This chapter aims to understand the benefits and costs of FDI from a host country perspective from economic and social welfare point of view. In this regard, some of the key benefits and costs of FDI are discussed with respect to employment generation, transferring technology and management resources, balance-of-payments, and FDI's impact on local competition. Chapter three presents an outlook of Bangladesh's power sector starting from the early years of 1994-2008 through the Sixth Five-Year Plan and the Seventh Five-Year Plan to the Eighth Five-Year

Plan for attaining the power generation capacity enhancement targets. This chapter also explores the different reform initiatives completed during this period with the enactment of the key policies, unbundling of the power sector, potential constraints and challenges in achieving the growth targets and modification of strategies in a realistic setting to achieve future development targets, and having universal access to power for all. Chapter five depicts the key factors that encourage foreign investors to conduct FDI in the Bangladesh power sector which are subsumed under four broad categories of investment prospects, namely regulatory, economic and financial, political, and societal. This chapter also underscores the theoretical background of this study using Dunning's eclectic paradigm and a host country's institutional setup (i.e., the quality of institutions) under the four broad investment dimensions for attracting or deterring FDI in the power sector. Chapter six highlights the key barriers that deter FDI in Bangladesh's power sector. Chapter seven presents the conclusion and policy implications for sustainable FDI generation in the Bangladeshi power sector for both conventional sources (i.e., fossil-fuel) and renewable power generation. It also presents areas for future research as a logical follow-up for the future direction of attracting additional FDI in the power sector.

This book makes a valuable contribution to our knowledge about conducting FDI in the power sector both for conventional energy (i.e., coal, gas, and oil) and renewable power companies in a fast-growing developing country like Bangladesh which primarily relies on fossil fuels to sustain the high-energy intensity of its growing economy while at the same time partially opening up its market for variable renewables or intermittent power to sustain its economic growth trajectory. Moreover, this book considers a wide range of institutional variables/factors for conducting FDI in the power sector under four broad categories of investment prospects (i.e., regulatory, economic and financial, political, and social) and takes a wider sample frame (both conventional energy sources and renewable power companies) for informing investment decisions in the power sector. In this way it has contributed to the extant literature on FDI in the power sector of developing countries by moving beyond the regulatory mechanism of rules and regulations and investment incentives to conduct FDI, thus providing a holistic look and adding a wider lens to the investors repertoire when considering strategic investment decision in the power sector of emerging economies. The arguments put forward in this study will inform governments on how to improve their institutional setups for attracting greater FDI flows in their respective power sectors.

This book is primarily designed for academics, managers, and professionals in the power industry; foreign investors who are seeking to invest in the Bangladesh's power sector, especially looking from a wider range of investment perspective and have a composite view of the factors that help attract and deter FDI in the Bangladeshi power sector.

With best regards,
Tareq Mahbub

Contents

About the Author

Tareq Mahbub is a Professor at the Melville School of Business at Kwantlen Polytechnic University, British Columbia, Canada. He completed his PhD in Management from the renowned Asian Institute of Technology (AIT) Bangkok and MBA in International Business from Carleton University, Ottawa. His areas of interest are foreign direct investment (FDI), FDI and intellectual property rights, international market entry modes and FDI and renewable energy. He has over 10-years' experience in the pharmaceutical sector working as Business Development Manager for a top-notch US FDA and UK MHRA certified international pharmaceutical company in the areas of new market development, contract manufacturing, under-licensing and strategic alliances both in the regulated and moderately regulated markets.

Acronyms

ADP	Annual development program
APSCL	Ashuganj Power Station Company Limited
BERC	Bangladesh Energy Regulatory Commission
BEZA	Bangladesh Economic Zones Authority
BIDA	Bangladesh Investment Development Authority
BOI	Board of Investment
BPC	Bangladesh Petroleum Corporation
BPDB	Bangladesh Power Development Board
BREB	Bangladesh Rural Electrification Board
BWDB	Bangladesh Water Development Board
CBAs	Collective bargaining agents
CEI	Chief electrical inspector
CPTU	Central Procurement Technical Unit
CSP	Concentrated solar power
CVE	Countering violent extremism
DESA	Dhaka Electric Supply Authority
DOE	Department of Environment
DPDC	Dhaka Power Distribution Company Limited
DTAA	Double Taxation Avoidance Agreement
ECA	Export Credit Agency
ECC	Environment Clearance Certificate
EGCB	Electricity Generation Company of Bangladesh
e-GP	e-government procurement
EIA	Environmental impact assessment
EPC	Engineering, procurement and construction
ESIA	Environmental and social impact assessment
FDI	Foreign direct investment
FIT	Feed-in-tariff
FSA	Fuel Supply Agreement
FSRU	Floating storage regasification unit
FY	Financial year

FY	Fiscal year
G2G	Government-to-government
GCEF	Global Community Engagement and Resilience Fund
GDF	Gas Development Fund
GDP	Gross domestic product
GHG	Greenhouse gas
GWh	Gigawatt hour
HFO	Heavy fuel oil
ICC	International Chamber of Commerce
ICCB	International Chamber of Commerce Bangladesh
ICSID	International Center for Settlement of Investment Disputes
ICT	Information and communications technology
IDCOL	Infrastructure Development Company Limited
IFC	International Finance Corporation
IOCs	International oil companies
IPPs	Independent power producers
IPR	Intellectual property rights
ITLOS	International Tribunal for the Law of the Sea
JV	Joint venture
KV	Kilovolt
kWh	Kilowatt hour
LDC	Least developed country
LGED	Local Government Engineering Department
LNG	Liquefied natural gas
M&A	Mergers and acquisitions
MMPTA	Million ton per annum
MMSCFD	Million standard cubic feet per day
MNE	Multinational enterprise
MW	Megawatt
NBR	National Board of Revenue
NDC	Nationally determined contributions
NWPGCL	North-West Power Generation Company Limited
O&M	Operation and maintenance
OECD	Organization for Economic Co-operation and Development
P&G	Proctor and Gamble
PBS	Palli Bidyut Samity
PGCB	Power Grid Company of Bangladesh
PP	Perspective plan
PPP	Power Purchase agreement
PPR	Public procurement reform
PPRP-II	Public Procurement Reform Project-II
PSC	Production sharing contract
PSMP	Power sector master plan
PSRB	Power sector reform in Bangladesh
PV	Photovoltaics

R&D	Research and development
REB	Rural electrification board
RFP	Request for proposal
RHD	Roads and Highways Department
SBU	Strategic business unit
SHS	Solar home system
SIPP	Small independent power producer
SREDA	Sustainable and Renewable Energy Development Authority
SSms	Super-super markets
T&D	Transmission and distribution
TCF	Trillion cubic feet
TWP	Temporary work permit
UK	United Kingdom
UN	United Nations
UNCITRAL	United Nations Commission on International Trade Law
UNCTAD	United Nations Conference on Trade and Development
US	United States
WIPO	World Intellectual Property Organization
WTO	World Trade Organization
WZDPC	West-Zone Power Distribution Company Limited

Chapter 1
Introduction

The United Nations explains energy is the key to every new opportunity and challenges the world faces today. This is for jobs, security, climate change, food production or increasing incomes – access to energy for all is essential. It has been the fact that about one billion people living in Sub-Saharan Africa and South Asia have no access to electricity. This proves to be a significant barrier to progress for a sizable population of the world and has impacts on a wide range of development indicators including basic education, food security, gender equality, livelihood and poverty reduction. Though the number of people getting access to power has improved since 2010 which is estimated around 118 million per year, however, these efforts need to accelerate at a considerable pace if the world is to meet the United Nations Sustainable Development Goal 7 – that is ensuring affordable, reliable, sustainable and modern energy for all by 2030. Some of the impediments that are confronting developing countries for having universal access to electricity are a lack of sufficient power generation capacity, poor transmission and distribution infrastructure, high-cost of supply to remote areas or a lack of affordability for electricity which are some of the leading issues for having access to grid-based electricity. However, on a positive note a number of countries have made considerable progress for access to electricity using a mix of techniques for example, Bangladesh has used both its off-grid solar home system and an extension of the main grid to increase the proportion of people to access to electricity from 32% to 62% between 2000 and 2014; China and India both focusing on-grid and off-grid solutions, Vietnam and Ghana primarily focusing on on-grid electricity and Kenya has lighted up 700,000 homes through its home solar system with its widely popular pay-as-you-go model widely popular in Sub-Saharan Africa.

In this context Bangladesh is a fast-growing developing country in South Asia. It is part of 11 emerging economies. Bangladesh's current government has set 'Vision 2021' to move Bangladesh to an upper middle-income country by the year 2021. The plan also stipulates to have universal access to electricity by 2021. In line with

T. Mahbub, *Encouraging Foreign Direct Investment (FDI) in Bangladesh's Power Sector*, SpringerBriefs in Energy, https://doi.org/10.1007/978-3-031-27990-4_1

this strategic vision, it has undertaken large programs for infrastructure development in partnership with private sector including foreign direct investment (FDI), in particular in the power and energy sector. In Bangladesh the investment requirements for power generation beyond 2015 are estimated at US$7.5 billion by 2021 and a further US$70 billion by 2035. Demand for electricity is growing at 8% a year. More importantly, 93% of the population currently has access to electricity [1, 2]. Accordingly, the government has embarked upon a comprehensive energy development strategy. This include new capital outlays in generation, transmission, and distribution; diversifying the energy mix from gas to cleaner coal technologies, liquefied natural gas (LNG) imports, energy trade, renewable energy development, conserving energy, and other better use of existing installed power generation capacities. As meeting these large investment requirements would exceed the ability of the public sector and the government has to allocate its limited resources to other competing priorities including the social sector, much of these investments would be conducted through private and foreign capital including FDI. The government has also scaled up its budget for the power sector and has embarked on a wide-scale institutional and sectoral reform programme to attract FDI. For example, the allocation for power and energy[1] sector has been raised by Taka[2] 726 crores[3] in the budget of fiscal year (FY) 2021–2022 which is Taka 27,484 up from Taka 26,758 crore in FY 2020–2021. Some of the reforms in progress are (i) adoption of short-term, mid-term and long-term plan for capacity expansion from 2016–2030; (ii) reliance on more large coal based power plants; (iii) phasing out of the quick rental power plants to reduce the power generation cost and subsidy for imported fuel; (iv) renewable energy development; (v) installation of country wide uniform prepayment metering system for preventing system losses and increasing revenue collection and reduce account receivables; (vi) pricing policies regarding increasing bulk tariffs and retail tariffs in a gradual manner; (vii) improving transparency and accountability in the sector; (viii) computerized billing; and finally (ix) setting up a dedicated state owned entity to implement and monitor large-coal based power plants in the country.

To realize this vision, Bangladesh government has instituted the power sector master plan (PSMP) 2010 which has been subsequently followed by PSMP 2016 subsumed under the 6th Five-Year Plan and the 7th Five-Year Plan respectively. The PSMP 2016 indicates that to attain a 8% GDP growth and to have universal power for all by 2021, the installed generation capacity target would be 23,000 megawatt (MW) by 2020 MW, 24,000 MW by 2021 and 40,000 MW by 2030 [3]. It has been estimated that Bangladesh would require US$40 billion investment for new generation by 2021 and another US$ 42 billion for new generation, distribution and

---

[1] Generally energy sector includes both renewable and non-renewable energy sources (e.g., fossil fuels, gas) that are used to produce electricity in power plants. Power generation is considered a sub-sector of the energy sector. In Bangladesh the energy sector comprises both renewable and non-renewable energy sources (e.g., gas and petroleum) in power generation.

[2] The Taka is the official national currency of Bangladesh

[3] One crore amounts to ten million

transmission by 2030 (The Daily Star, 2016) to meet the constant demand supply gap that has been widening over the years [4]. and its growing energy needs which is set on a path of rapid industrialization based on export-led growth. Especially for power sector capacity expansion and enhancing the energy efficiency and intensity along with rising economic and population growth and associated rising incomes and shifts from rural to urban areas.

The power and energy sector are the two sectors in Bangladesh for which FDI has been encouraged through the provision of various support policies. Some of these are: (i) tax exemption on royalties, technical assistance fees and facilities for their repatriation; (ii) tax exemption interest on foreign loans; (iii) tax exemption on capital gains from transfer of shares by the investing company; (iv) avoidance of double taxation; (v) exemption from income tax for up to 3 years for expatriate personnel employed under the approved industry; (vi) remittance of up to 50 percent of the salary of foreigners employed in Bangladesh and facilities for repatriation of their savings and retirement benefits at the time of departure; (vii) no restriction on issuance of work granted to project-related foreign nationals and employees; and (viii) facilities for repatriation of invested capital, profits and dividends.

Despite such supporting measures and incentives for FDI, FDI in the power sector is not very encouraging. For example, FDI in the power sector increased from US$30 million in 2004–2005, accounting for 4% of the total FDI flows to US$ 208 million (10% of total FDI flows) and US$ 520 million (22% of total FDI flows) in 2019–2020. In 2019–2020, the share of FDI flows to the power sector surpassed the other major sectors of the economy, including textile and clothing, banking, and telecommunications [5] (Fig. 1.1).

However, it could be argued that much of the rise in FDI in the power sector in recent times could be attributed to the large inflow of capital from

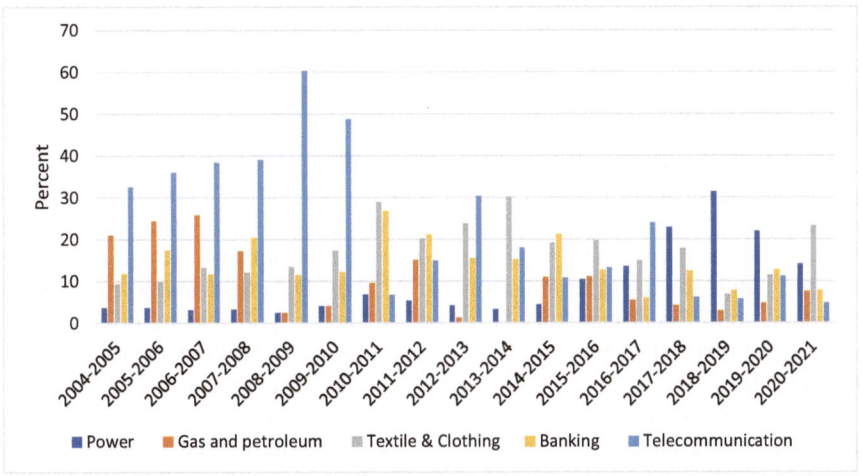

Fig. 1.1 Proportion of FDI in the Bangladeshi power sector, 2004–2020. (Source: Ref. [5])

Government-to-Government (G2G) FDI projects which is less attracted to country specific factors including various incentives and support measures to attract FDI if not for private invested capital to attract FDI but on bilateral relationships between countries.

Moreover, it has been argued that even with the introduction of the first private generation policy act in 1994 which delineates the rules of participation for individual power producers (IPPs) in Bangladesh including the various facilities and incentives for foreign investor participation in the power sector, no substantive increase in private sector participation in FDI has occurred in capacity addition except in 2015 in the installation of three large base-load gas power plants namely, Summit Meghnaghat 335 MW, Summit Bibiyana II 341 MW other than the provision of short-term rentals and expensive diesel or Heavy Fuel Oil (HFO) IPPs.

Therefore, based on the above background it would be more interesting to investigate the factors that attract FDI in the Bangladesh power sector. Especially identifying the factors from the four aspects of investment perspectives such as regulatory, political, economic and financial and social point of view that are deemed priorities from foreign investors to undertake FDI in the power sector. It is assumed that identifying these factors through an empirical lens would provide insights into the perceptions of foreign investors into the factors that are deemed important in their investment decision-making into investing in the Bangladesh power sector and informing the government as to picking up the right signals for creating an enabling environment for sustainability of FDI for the future.

There are numerous studies that examine FDI determinants in the general FDI literature mainly focusing on FDI flows to developing countries. The determinants empirically tested can be categorized into three dimensions namely, macroeconomic environment, institutional environment and natural conditions [6]. Some studies argue that market size, infrastructure, labor cost and exchange, trade openness and tax structure attract or deter FDI in developing countries [7–9]. In the institutional environment dimension factors such as political risk, rule of law (effective law enforcement), corruption and efficient and transparent administrative procedure have both positive and negative correlations in attracting or deterring [10–12]. In the natural condition dimension for example, acquisition to land has been a major determinant in attracting FDI [13].

To draw the distinction between the general determinants of FDI and FDI in the power sector, it could be argued that most of the general literature on the location choice of FDI especially in the developing countries are based on aggregated FDI but lack detailed sector-analyses data or the FDI determinants that are relevant to specific industries for FDI attractiveness. As the FDI attractiveness could differ according to different industries based on the support policies and incentives for specific industries (i.e., power), for the location choice of Multinational Enterprises (MNEs). Therefore, there is a need to investigate the location choice of FDI in specific industries. There is a need to view FDI (i) shifting the focus from aggregated FDI (general FDI) to FDI by sector; and (ii) the need to have more empirical studies on support policies on the location choice of FDI on a sector-basis. FDI in the power sector is different in the sense that it is more influenced by the regulatory factors/

environment in the host country and policy interventions from country-specific, sectoral level and project specific dimensions subsumed under the role of institutions that draw FDI in this sector. This is quite different from other general FDI literature that focus on the country-specific dimension as a whole rather than sector-specific. However, when analyzing FDI in the power sector along with the regulatory factors that include specific policies and incentives at the sectoral level, the overall institutions of the country with variables like corruption, political risk, labor cost, natural resources and economic growth and development need to be considered.

Most of the FDI literature in the power sector for developing countries primarily look into the regulatory aspects of host countries in the location choice for conducting FDI and takes a cursory look into other institutional factors that attract FDI in this regard. Therefore, the author has attempted to take a wider look into the factors categorized into the four broad areas of investment prospects and a wider range of variables to have a holistic look from foreign investors perspective and the set issues that are important to conduct FDI in the Bangladeshi power sector.

The book is organized as follows. The first chapter presents an introduction of the background of this study especially Bangladesh's vision to become a middle-income country by 2021 and have universal power for all. It also depicts how power is shown a principal catalyst for increasing Bangladesh's GDP growth set on a path of rapid industrialization and export oriented dynamic economic growth. It also depicts the rationale for the study especially how FDI is seen as a catalyst in achieving its future development objectives. Chapter 2 presents an overview of some of the benefits and costs of FDI for a host nation from an economic and social welfare point of view. Chapter 3 presents an outlook of Bangladesh's power sector starting from the early years of 1994–2008 through the Sixth Five-Year Plan (2011–2015) and the Seventh Five-Year Plan (2016–2020) to the Eighth Five-Year Plan for attaining the power generation capacity enhancement targets. Chapter 4 depicts the key factors that encourage foreign investors to conduct FDI in the Bangladeshi power sector which are subsumed under four broad categories of investment prospects namely, regulatory, economic and financial, political and social. Chapter 5 highlights the key barriers that deter FDI in Bangladesh's power sector both for conventional fossil-fuel power companies and renewable energy. Chapter 6 presents the conclusion and policy implications for sustainable generation of FDI in the Bangladesh's power sector. This chapter also depicts the areas for future research for sustainability of FDI in the power sector.

References

1. CPD (2021) Bangladesh economy in FY 2020–21: interim review of macroeconomic performance. https://think-asia.org/bitstream/handle/11540/14088/Bangladesh-Economy-in-FY2020-21-Interim-Review-of-Macroeconomic-Performance.pdf?sequence=1. Accessed 8 Sept 2022
2. GOB (2020) 8th Five Year Plan July 2020 – June 2025: promoting prosperity and fostering inclusiveness. https://policy.asiapacificenergy.org/sites/default/files/Eighth%20Five%20Year%20Plan%20%28EN%29.pdf. Accessed 7 Sept 2022

3. GOB (2015) 7TH Five Year Plan FY 2016- FY 2020: accelerating growth, empowering citizens. https://www.unicef.org/bangladesh/sites/unicef.org.bangladesh/files/2018-10/7th_FYP_18_02_2016.pdf. Accessed 5 Oct 2022
4. USAID (2021) Recommendation for a renewable energy implementation action plan for Bangladesh. https://pdf.usaid.gov/pdf_docs/PA00XD5J.pdf. Accessed 2 Aug 2022
5. Bangladesh Bank (2020) Foreign direct investment and external debt, 2020. https://www.bb.org.bd/pub/halfyearly/fdisurvey/fdisurveyjuldec2020.pdf. Accessed 3 Jan 2022
6. Keeley RA, Matsumoto K (2018) Investors' perspective on determinants of foreign direct investment in wind and solar energy in developing economies – review and expert opinions. J Clean Prod 179:132–142
7. Shamsuddin AF (1994) Economic determinants of foreign direct investment in less developed countries. Pakistan Dev Rev 33(1):41–51
8. Contractor FJ, Nuruzzaman N, Dangol R, Raghunath S (2021) How FDI inflows to emerging markets are influenced by country regulatory factors: an exploratory study. 27:100834
9. Uddin M, Boateng A (2011) Explaining the trends in the UK cross-border mergers and acquisitions: an analysis of macro-economic factors. 20(5):547–556
10. Alfaro L, Kalemli-Ozcan S, Volosovych V (2008) Why doesn't capital flow from rich to poor countries? An empirical investigation. Rev Econ Stat 90:347–368
11. Busse M, Hefeker C (2007) Political risk, institutions and foreign direct investment Eur. J Polit Econ 23:397–415
12. Edwards S (1990) Capital flows, foreign direct investment, and debt-equity swaps in developing countries NBER Cambridge working paper 3497
13. Escaleras M, Register CA (2011) Natural disasters and foreign direct investment. Land Econ 87:346–363

Chapter 2
Benefits and Costs of FDI

Abstract This chapter gives an overview of the some of the benefits and costs of FDI from a host country perspective. Some of the aspects which are discussed here are (i) employment effects of FDI; (ii) resource transfer effects; (iii) balance-of-payments effects; and (iv) effects of competition and economic growth.

Keywords FDI · Competition · MNE · Technology · Spillovers

2.1 Employment Effects of FDI

One beneficial effect of FDI is to create jobs in the local economy. This effect of FDI is both direct and indirect. The direct effect rises when an MNE directly recruits local people from the host country for its operations. Indirect effect refers to creating auxiliary jobs in the local suppler base for direct presence of the MNE in the local market with respect to increased local spending by employees by the MNE. It is argued that indirect employment effects are not as large as the direct effects. One consideration is when Toyota launched its newest auto plant in France in 1998 it was suggested that the plant would create 2000 new direct jobs and another 2000 jobs in the supporting industries [1]. In a similar fashion Toyota has invested US$ 51 billion in manufacturing in the United States (US) over the last several decades and have created 94,000 manufacturing jobs in their US plants and 1.6 million indirect jobs in the industry by way of dealerships and suppliers in the US [2].

However, there are arguments that suggest that some of the beneficial effects of new job creation in the local market could be offset by job losses in the local industry. For example, it has been suggested that the jobs created by FDIs in Japanese auto companies in the US has been more than offset by job losses in the local auto industry, as US companies lost market share against Japanese competition. Therefore, any net gain of employment conducted by FDI for a specific industry in the host country in the initial period of market entry could be a major negotiating point between the MNE and the host country government. Evidence suggests such negative effects (initial job losses) in the local industry by competition from FDI become offset my more employment creation (the net number of new jobs) at a faster rate in the later years [1]. For example, when FDI takes the form of acquisition in a host country evidence suggests that there are initial layoffs for three main reasons. These are: rationalizing, enhancing efficiency (particularly privatised companies) and reducing excess capacity. However, in the long-term if the restructuring is successful the MNE tends to increase their employment base in the local economy at a faster rate than domestic rivals.

2.2 Resource Transfer Effects of FDI

Foreign firms can make a positive contribution to a host country by supplying capital, technology and management resources that may not be otherwise available in the local economy and boost the country's economic growth rate. With respect to capital MNEs come with large financial strengths with respect to sourcing of capital through their internal company sources or because of their reputation are in the standpoint to borrowing large amounts of capital from capital markets than their local counterparts in the host country. It has been suggested that technology can stimulate economic development and industrialization. Technology can take two forms. For example, the technology can be incorporated into a production process (i.e., the technology for discovering, extracting and refining oil) or it can be built into a product (i.e., personal computers) [1]. Evidence suggests that many countries lack the research and development (R&D) resources and skills to develop their own indigenous product and process technologies and FDI could be a means to provide such resources.

Research suggests that MNEs often transfer significant technology when they invest in a host country. For example, a study conducted in Kenya's energy sector from the period 2001–2014 on the role of FDI in technology transfer and economic growth found that the presence of foreign firms led to the development of efficient procurement networks for production and sales of goods internationally, transferring technology and establishing markets for domestic production, increased domestic savings and improved investment policies [3]. Another study by Organization for economic co-operation and development (OECD) found that foreign investors invested significant amounts of capital in R&D in the countries in which they invested which suggests that these MNEs were not only transferring technologies to the host countries but also upgrading existing technology and creating new technology in the process [1]. However, it has been argued that for firms in developing countries to benefit from the resource transfer effects of FDI is part of the organizational learning process through linking, leveraging and learning capture to enable their technological development. For indigenous firms the process starts in importing new technology and then investing in building their capabilities to master the tacit[1] elements. As firms draw on internal and external linkages (both foreign and domestic) to build their capabilities. The process starts with capabilities needed to master the technology for production purposes which may deepen over time to improving the technology and creating a new technology. This is also a function of the level of investments by indigenous firms as to how much they need to benefit from the resource transfer effects of FDI which is moderated by the competition faced in foreign and domestic markets as well as on their abilities to access complementary supporting activities (i.e., supply-chain linkages) [3].

Another important benefit for FDI in the host country is related to technology spillovers. It is suggested that spillovers are indirect effects of inward FDI and are considered a function of an MNE's unintended transmission of knowledge and skills from the FDI enterprise to domestic firms through demonstration effects (i.e., imitation, reverse engineering and R&D) and/or workers' mobility [4, 5]. Since it is assumed foreign firms produce a higher level of technology than local firms; therefore they can stimulate such effects. Spillovers are also dependent on the differences on the level of technological intensity between foreign firms and local firms and the degree of export orientation of foreign firms. Counterintuitively it has been suggested that if a foreign firm lacks enough external linkages in the local economy it is less likely to generate additional spillovers to local firms.

There is another important benefit that accrues for local personnel working for foreign firms in the host country's economy. For example, local employees can also learn the technology while working for the firm and some of them may leave the foreign firm to help establish indigenous firms using the acquired technology from the MNE [6].

Though it has been suggested that FDI may have wider technological benefits through its spillover effects, countervailing evidence also suggests that FDI could also discourage the development of technical know-how by and in local firms and to the detriment of the growth of domestic producers and the national economy. Other factors on the cost side are social costs in the form of unemployment when FDI, which is relatively capital intensive causes the more labor-intensive local firms to close down [7]. It could be argued at this point that until the 1980s Japan had been very restrictive toward inward FDI based on the assumption that Western FDI with their ample managerial knowledge and superior technological know-how would restrict the growth and development of its own indigenous technologies. For the local knowledge-base and indigenous technologies to develop and flourish, Japanese government discouraged local firms to get into joint ventures (JVs) or licensing arrangements with foreign firms based on the premise that such fledging and nascent technologies need to be protected from undue foreign intervention to allow for their unimpeded development and growth. However, counterintuitively it was also found that for Japanese companies to compete internationally the Japanese government developed policies that would allow Japanese companies to license foreign

---

[1] Tacit knowledge refers to knowledge, skills and abilities an individual gains through experience that is difficult to put into words or otherwise communicate.

technologies and persuaded foreign firms to share their intellectual properties with Japanese firms while subsidizing research and development costs for Japanese firms that intended to build their products based on foreign technologies. This helped to protect the local market from direct foreign competition. However, foreign firms which had important technology that could impart significant benefits to the Japanese economy – the government encouraged such new technologies to be diffused in the Japanese economy provided that these firms do not enter into any forms of licensing or JV opportunities with Japanese firms. Based upon this stance IBM and Texas Instruments set up their own subsidiaries in Japan in the early to mid 1980s.

2.3 Balance of Payment Effects

FDI has both a both a positive and negative effect on a host country's balance of payment positions. A country's balance of payments tracks both its payments to and its receipts from other countries. Governments are normally concerned when their country is running a deficit in the current account of their balance of payments. The current account tracks the export and import of goods and services. There are two way an FDI can help improve a country's balance of payments position. First if FDI is a substitute for imports it can improve the current account of the host country's balance of payments as could be illustrated that much of the FDI by the Japanese automobile companies in the US and United Kingdom (UK) in their initial phase of market entry were chiefly looked upon as substitution for imports from Japan which improved the balance of payments positions for these two countries. However, counterintuitively it could be argued that it could have a negative effect on the balance of payments if the foreign subsidiary increases its import of inputs from the home country and repatriate large amounts of capital abroad. As these outflows of capital from the host country are shown as debit on the current account of the host country's balance of payment positions.

A second consideration is that could have a significant positive effect on a host country's balance of payments position from conducting FDI is that when the FDI is directed towards exports to third country markets or rich regional markets, these can substantially improve a host country's balance of payments. A classic example is Volkswagen which in 1992 acquired Skoda, the national car company of the Czech Republic and redirected its sales efforts to the European Union which resulted in an export boom for the Czech economy [7, 8]. This help foreign subsidiaries to mobilize their resources both in the host economy and in the global markets thus earning significant foreign exchange, improving the country's balance of payments position, creating more local jobs and contributing to the country's economic growth.

2.4 Effect on Competition and Economic Growth

In a market economy firms compete with each other to win customers. Competition provides incentives for firms to increase productive efficiency, producing high quality products and service, increase the variety of products to consumers and ensure competitive pricing. Competition leads domestic firms to engage in innovation, aiming at achieving higher efficiency in production through differentiation, quality improvement and technological advancement [4]. It is argued that local firms benefit from competition through the arrival of foreign firms when they possess the capability to absorb technology and higher skills (Li, 2008) [9]. It is suggested that when FDI takes the form of greenfield investment, it encourages new enterprises increasing the number of players in the market giving consumers more choice in products and services. More importantly, increased competition encourages foreign firms to invest in plant, equipment and R&D to get a competitive advantage over their domestic rivals. This helps for increased productivity growth and product and process innovations. For example, when South Korea liberalized its retail sector in 1996 to encourage competition from large Western discount stores like Walmart, Costco, Carrefour and Tesco – this seems have encouraged local discounters to improve their efficiency and compete with these large discount retailers. Against the backdrop of their arrival though small business owners and traditional markets lost market share in increased competition but large local retailers and super-super markets (SSms) gained additional market share through restructuring their operations against the competition. Currently, the large retail discount stores have been consolidated into Korean Big 3 (E-mart, Homeplus and Lotte-Mart) which holds a market share of 80.5% in South Korea [10].

However, FDI could also come with some costs. These are (i) adverse effects on competition; (ii) adverse effects on the balance of payments; and (iii) effects on national sovereignty and autonomy on a host country. These are discussed below.

2.5 Adverse Effects on Competition

This concept holds that host governments sometimes have concern that the competition generated by the presence of an MNE in the host country can have detrimental effects for local firms. For example, MNEs have greater economic power in terms of their ability to draw on funds generated elsewhere to subsidize its costs in the host market. This has the effects of driving out local firms out of business and allow the MNE to monopolize the market. FDI typically results in increased competition in both output markets and markets for skilled labor and other inputs which may result in the crowding out of domestic firms. It has been argued that once the market is monopolized the foreign MNE can raise prices above those that would prevail in competitive markets with harmful effects on the economic welfare of the host nation. This effect tends to be greater in less developed countries that have few larger firms of their own as opposed to large industrialized countries that have a larger concentration of both competing local and foreign firms. Looking at the horizontal impact of inward FDI (i.e., the firm duplicates its business activities in the home country into other countries) the expected effect on domestic competition is largely negative as the presence of foreign competition not only adds to increased competition for labor and other inputs by domestic firms but also reduces output prices with the effect that both the input and output sides raise the exit hazard for domestic firms [11]. In the context of vertical FDI (i.e., when a firm invests internationally to provide inputs into its core operations – usually in its home country) the effect of competition is positive as it generates various productivity spillovers through the increase in product varieties and the use of specialized inputs from backward linkages or the technical support and guidance provided through forward linkages. However, negative effects could result when FDI suggests changes in technological standards and quality requirements, especially in upstream sectors (i.e., oil and gas) for competing local firms downstream that supply inputs with local prices and quality forcing local firms to adjust technologies.

2.6   Adverse Effects on the Balance of Payments

The adverse effects on the balance of payments position on a host country are twofold. First, there is an inflow of funds that comes with an FDI which is subsequently balanced with an outflow of earnings from the foreign subsidiary to the parent country. Such outflows are shown as capital outflows on the balance of payments accounts. Governments sometimes attempts to restrict such outflows by putting a cap as to how much earnings that can be repatriated to a foreign subsidiary's home country. A second consideration is when a subsidiary imports a large number of inputs from abroad which is shown as a debit on the current account of the host country's balance of payments. This effect could be offset when the foreign subsidiary starts to use local contents or domestically produced parts and components in its manufacturing of products. For example, one critique that could be labelled for the Japanese owned auto assembly plants in their initial foray into the US was that these firms tended to import many component parts from Japan. However, later this has been balanced as Japanese auto companies started operating in the US by pledging to use 75% of their component parts from US based manufacturers if not US owned manufacturers. When Nissan invested in UK it responded to concerns about local content by pledging to increase the proportion of local content to 60 percent and then to 80 percent respectively [1].

2.7   Effects on National Sovereignty and Autonomy of Host Countries

There have been some concerns from host governments that FDI is accompanied by some of loss of economic freedom. One consideration is that the key decisions that affect the country's economy could be made by foreign affiliates that have no real commitment to the host economy over which the state has no real control. However, this assumption has been dismissed by economists as a product of outmoded thinking as it fails to account for the growing interdependence of the world economy. It has been argued that in a globalized world where national economies are merging into an interdependent and integrated global economic system and are investing into each other's markets, therefore, one country holding another country to "economic ransom' will be detrimental to itself own development and growth. In this regard it could be argued that some of the giant multinationals like Apple, Microsoft, Unilever, Nestle and Procter and Gamble (P&G) in their size and market capitalization are even larger than many small host country's GDP. These multinationals have the ability to exercise considerable influence on host government's economic policies for their ability to transfer resources for example, capital, technology and management skills for boosting host countries' economic development. Thus, creating local jobs, increase productive efficiency both from competition from local and foreign companies, improvement in the quality of the factors of production, lowering prices of goods and the introduction of new or better-quality goods including giving small host economies access to a wider international export network for goods and services which otherwise the host countries cannot develop on their own.

Countervailing arguments also suggest that imposing burdensome restriction on FDI could inspire retaliatory policies from other nations. On this basis, in order to avoid such measures, the Organization for Economic Development and Cooperation (OECD) as well its 12 nonmember states have signed a nonbinding commitment to treat foreign firms receiving the same treatment in their territories as with any local firms. These governments have identified areas for example, physical and intangible assets or critical infrastructure whose destruction or disruption would seriously undermine public safety, social order and the fulfillment of key government responsibilities that warrant protection from the states. Likewise, foreign investment is restricted in these sectors for grounds related to national security. Some of these critical infrastructures are: energy (including nuclear), information and communications technology (ICT), finance, healthcare, food, water, transport, chemicals and defense industry base.

The following table gives an overview of some of the definitions of critical infrastructure in the listed countries which tend to be broad. The definition of 'critical infrastructure' refers to infrastructure whose disruption or destruction would cause catastrophic and far reaching damage for the country concerned (Table 2.1).

There is another interesting argument that is worth mentioning in this regard. Many host countries have raised concerns to inward FDI especially developed countries and have resorted to hostile measure to block such investments. It has been

Table 2.1 National differences in critical infrastructure

Australia	"Critical infrastructure is defined as those physical facilities, supply chains, information technologies and communication networks which, if destroyed, degraded or rendered unavailable for an extended period, would significantly impact on the social or economic well-being of the nation, or affect Australia's ability to conduct national defence and ensure national security."
Canada	"Canada's critical infrastructure consists of those physical and information technology facilities, networks, services and assets which, if disrupted or destroyed, would have a serious impact on the health, safety, security or economic well-being of Canadians or the effective functioning of governments in Canada."
Germany	"Critical infrastructures are organisations and facilities of major importance to the community whose failure or impairment would cause a sustained shortage of supplies, significant disruptions to public order or other dramatic consequences."
Netherlands	"Critical infrastructure refers to products, services and the accompanying processes that, in the event of disruption or failure, could cause major social disturbance. This could be in the form of tremendous casualties and severe economic damage…"
United Kingdom	"The [Critical National Infrastructure] comprises those assets, services and systems that support the economic, political and social life of the UK whose importance is such that loss could: (1) cause large-scale loss of life, (2) have a serious impact on the national economy; (3) have other grave social consequences for the community; or (4) be of immediate concern to the national government."
United States	The general definition of critical infrastructure in the overall US critical infrastructure plan is: "systems and assets, whether physical or virtual, so vital to the United States that the incapacity or destruction of such systems and assets would have a debilitating impact on security, national economic security, national public health or safety, or any combination of those matters." For investment policy purposes, the definition is narrower: "systems and assets, whether physical or virtual, so vital to the United States that the incapacity or destruction of such systems and assets would have a debilitating impact on national security."

Source: Ref. [12]

suggested that these reactions have come for foreign transactions that are typically associated with mergers and acquisitions (M&As) rather than investments by green-field ventures [13]. For example, US lawmakers along with their peers have passed legislation that restricts or authorizes reviews for foreign deals that could have significant impacts on the economy such as outsourcing of jobs, sharing sensitive technologies or impairment of critical infrastructure. This is an attempt to protect such companies which are vital to national interests and host governments typically do not want such strategic assets falling prey to foreign companies. In 2005 China National Offshore Oil Company had to withdraw a takeover bid for UNOCAL following a highly negative reaction from in Congress about the proposed takeover of a 'strategic asset' by a Chinese company. In the most recent times The Trump Administration had blocked two takeovers (i.e., September 2017 and in March 2018) by Chinese companies for example, the sale of the chipmaker Lattice Semiconductor in a deal partially financed by Chinese state-owned capital and the proposed takeover US telecom leader Qualcomm by a Singapore-based company [14]. The rationale for these actions were to protect US technological competitiveness and not to impair notational security. Similar instances have happened globally by other foreign nations for blocking Chinese acquisition activity in Europe and Australia. In 2016 Australia rejected a Chinese bid to buy the country's largest agribusiness and subsequently blocked a Chinese consortium from acquiring an electric grid operator. In Germany, China's top investment destination in Europe raised concerns about Chinese investments in acquiring German companies, for example Daimler stepped up its efforts to review and block foreign deals in 2018. Similarly, France and UK have increased regulators' authority and requiring state approvals for most foreign bid [12].

2.8   Summary

This chapter summarizes some of the benefits and costs of FDI from a host nation's perspective. In this regard some of the benefits and costs of FDI with respect to economic and social welfare are mentioned. These are (i) employment effects of FDI; (ii) resource transfer effects; (iii) balance-of-payments effects and (iv) effects of competition and economic growth. From the host-country costs these are: (i) adverse effects on competition; (ii) adverse effects on the balance of payments; and (iii) effects on national sovereignty and autonomy. The chapter discusses some of the positive and negative arguments about FDI from both the home and host country viewpoints.

References

1. Hill CWL, Hult GTM (2019) International business: competing in the global marketplace. McGraw-Hill Education
2. CNBC (2009) Japanese automakers tout US-based jobs at all-time high as trump ramps up trade war. Accessed 11 Sept 2022
3. Osano HM, Kaine PW, Kaine PW (2016) Role of foreign direct investment on technology transfer and economic growth in Kenya: a case of the energy sector. J Innov Entrepreneurship 5:31
4. Esquivias MA, Harianto SK (2020) Does competition and foreign investment spur industrial efficiency?: firm-level evidence from Indonesia. https://www.sciencedirect.com/science/article/pii/S2405844020313384. Accessed 5 Sept 2022
5. Landman M, Ojanpera S, Kinsella SO, Clery N (2022) The role of relatedness and strategic linkages between domestic and MNE sectors in regional branching and resilience. J Technol Transf. https://link.springer.com/article/10.1007/s10961-022-09930-4. Accessed 5 Sept 2022
6. Chia SY (1997) Singapore advanced production base and smart hub of the electronics industry. Multinationals East Asian Integr. https://www.idrc.ca/sites/default/files/openebooks/806-6/index.html. Accessed 4 Sept 2022
7. UNCTAD (1997) Trade and development report 1997: globalization, distribution and growth. http://r0.unctad.org/en/pub/pdfs/tdr17ove.pdf. Accessed 8 Nov 2022
8. Hill CWL (2003) International business: competing in the global marketplace. Irwin McGraw-Hill Education
9. Le T (2008) Brain drain or brain circulation: evidence from OECD's international migration and R&D spillovers. Scottish J Polit Econ 55:618–636

10. Woohyoung K, Hallosworth AG (2015) Tesco in Korea: regulation and retail change. J Econ Human Geogr 107(3):270–281
11. Kokko A, Thong TT (2014) Foreign direct investment and the survival of domestic private firms in Vietnam. Asian Dev Rev 31(1):53–91
12. OECD (2008) Protection of critical infrastructures on the role of investment policies relating to national security. https://www.oecd.org/daf/inv/investment-policy/40700392.pdf. Accessed 7 Sept 2022
13. Masters J, McBride J (2018) Foreign investment and U.S. national security. https://www.cfr.org/backgrounder/foreign-investment-and-us-national-security. Accessed 6 Sept 2022
14. Baker L (2017) Trump bars Chinese-backed firm from buying U.S. chipmaker Lattice. https://www.reuters.com/article/us-lattice-m-a-canyonbridge-trump-idUSKCN1BO2ME. Accessed 6 Aug 2022

Chapter 3
An Overview of the Bangladesh Power Sector

Abstract This chapter systematically reviews the sectoral performance of the Bangladeshi power sector from post-independence starting from 1972 through post 2000 including the Sixth Five-Year Plan (2011–2015) and Seventh Five-Year Plan (2016–2020) in which the most significant improvements made to attain the government's Vision 2021 initiative to accede to a middle-income country status and providing universal power for all. It also reviews the different reform initiatives completed during this period with the enactment of different policies, sector unbundling and future strategies to attain the growth targets. It also analyses the potential constraints and challenges for achieving the growth targets from a country specific perspective and potential modifications of strategies in a realistic setting in line with the country's future economic development goals.

Keywords Power · IPP · BPDB · Demand · Installed capacity · PSMP

3.1 Post-Independence

Bangladesh accords a high priority to providing quality electricity to its population. After its independence from Pakistan in 1971 the first state-run power development company The Bangladesh Power Development Board (BPDB) was established in 1972 which was responsible for the planning, generation, and transmission of electricity. Bangladesh's electricity reform started in 1977 with the creation of the Rural Electrification Board (REB) making the provision of rural access to electricity as a fundamental goal of state policy. In its early years, the access to rural electricity grew slowly. By 2000, only 20% of the rural population had access to electricity. It is to be suggested Bangladesh Rural Electricity Board (BREB) had been the pioneer in Bangladesh's Solar Home System Program (SHS) – the largest national program in the world for off-grid electrification with the goal of reaching 4.1 million households by 2021 connecting 20 million people and contributing to 14% of coverage [1]. The Dhaka Electric Supply Authority (DESA) was created in 1991 as part of the

© The Author(s), under exclusive license to Springer Nature Switzerland AG 2023
T. Mahbub, *Encouraging Foreign Direct Investment (FDI) in Bangladesh's Power Sector*, SpringerBriefs in Energy,
https://doi.org/10.1007/978-3-031-27990-4_3

continuing unbundling of the power sector. DESA took over the distribution responsibility from BPDB and the official policy was amended in 1992 for inclusion of the provision of private investment in the power sector. However, the unbundling process failed to live up to its expectation and in 1994 the government adopted the Power Sector Reforms in Bangladesh (PSRB) policy paper in consultation with its major multilateral development partners i.e., the World Bank and the Asian Development Bank. The PSRB formed the basis for the subsequent power sector reforms which were initiated during 1994–2008. Its recommendations were: unbundling the power sector entities and the establishment of an independent regulatory commission. This was then subsequently being followed by the establishment of the Power Cell in 1995 – an agency focused on power sector reform and contracting individual power producers (IPPs) for private-public power procurement.

In 1996 Bangladesh launched its first power generation policy which was later amended in 2004 that formed the core for introducing the first IPPs in the Bangladeshi power sector to allow private companies to participate into the sector. This policy opened the door for private and foreign direct investment FDI into the power sector, which culminated in the successful commissioning of two modern gas-fired combined cycle based-load power plant namely, AES Haripur 360 MW and AES Meghnaghat (450 MW) in 2002 respectively crediting Bangladesh as a world leader in attracting private capital at some of the lowest power tariffs. At this time a further step of power sector unbundling was initiated with the formation of Dhaka Electric Supply Company (DESCO), a public limited company responsible for generation to take over power distribution in certain parts of the capital city Dhaka not served by Dhaka Electric Supply Authority (DESA) as a function of reducing any distribution bottlenecks for supplying power in the capital city. Despite some of these significant reforms the period from 1994 to 1997 was marked by "inadequate power generation capacity, frequent and extended plant outages, and high distribution losses mainly due to pilferage of electricity" [2]. In 1998 the government made another attempt of reform to reorient the ministry of power, energy and mineral resources to create the Power Division – an agency within the ministry with the overall management of Bangladesh's power sector. At that time five new contracts with IPPs brought 565 MW of new additional power to the overall power generation capacity by 2001. It is to be mentioned that the generation capacity in 1995 amounted to 2900 MW, the available generation capacity was only 2130 MW indicating high levels of non-availability and outages. Bangladesh's per capital power generation in 1995 only amounted to 84 kilowatt-hour (kWh) compared with 303 kWh in India and 328 kWh in Pakistan at that time. Table 3.1 shows the installed capacity in MW available capacity and electricity generation from 1994 to 2008. It shows that the improvement in power generation capacity i.e., installed capacity, available capacity and power generation was marginal over the years and a significant increase in load shedding which show the poor performance of the power sector during this period.

Table 3.1 Installed capacity, available capacity, electricity generation and load shedding from 1994 to 2008

Year	Installed Capacity (MW)		Available Capacity MW	Electricity Generation (GWh)		Load shedding (MW)
	BPDB	IPP	BPDB + IPP	BPDB	IPP	
1994	2608	0	1881	9784		540
1995	2908	0	2133	10806		537
1996	2908	0	2105	11474		545
1997	2908	0	2148	11857		674
1998	3091	0	2320	12882		711
1999	3310	302	2850	13872	578	774
2000	3310	380	2665	14318	1244	536
2001	3331	685	3033	14828	2192	663
2002	3320	810	3217	14449	3771	367
2003	3420	1260	3428	12880	6298	468
2004	3420	1260	3592	13342	7478	694
2005	3735	1260	3720	14067	7939	770
2006	3895	1260	3782	15416	8286	1312
2007	3872	1330	3717	15494	8244	1345
2008	3814	3814	4130	16155	9138	2087

Source: Ref. [3]

3.2 Post-2000

The post-2000 was marked by the launch of the 'Vision and Policy Statement on Power Sector Reforms' drawn up in February 2000 which identified nine objectives with the core aim of providing electricity to the entire country by 2020, making the sector more efficient and financially viable, and increase competition including private sector involvement in the power sector. This was later reinforced by the perspective plan of Bangladesh 2010–2021: Making Vision 2021 a Reality to afford universal power by 2021 forming the guiding principle of meeting the development targets of the 6th and 7th Five-Year plans. In 2004 Bangladesh's Cabinet Committee on Economic Affairs adopted the power sector development program which further initiated a wide-scale reform of the energy sector aimed at unbundling the regional power companies into separate independent units with BPDB forming the holding company. This reorganization was aimed at further corporatizing and commercializing the sector with the regional power companies taking over BPDB's generation, transmission and distribution functions from being a 'single buyer' to a competitive multi-buyer system. As a follow through on some of these developments in 2006 the government drew a road map that included an implementation strategy for many of the goals adopted in 2002 vision and policy statement. It is during this time about 500,000 new connections were launched every year between 2000 and 2007 which resulted in doubling the access of power to the people since the start of the decade

[2]. It is during this time, Bangladesh made at least 500,000 new connections each year between 2000 and 2007 resulting in the doubling the access to power for the people since the start of the decade. Additionally, over 200,000 SHS was installed at that time initiating the process of a mass electrification of renewables for the rural households.

3.3 The Sixth Five Year Plan

In 2010 the government adopted the power sector master plan (PSMP) 2010 which called for an ambitious generation expansion plan for a threefold increase in power generation capacity over the next 5 years. The PSMP forecasts that the grid system demand combined with demand side management for 2015, 2020 and 2030 would be 10,283 MW, 17,3044 MW and 33, 708 MW respectively. The plan also stipulates the commissioning of a number of quick rental and rental power plants as immediate stopgap measures to meet the increased demand in the short run [4]. It is to be suggested that Bangladesh faced a significant power crisis in April 2009 with power shortages becoming the worst ever owing to very little capacity addition to the grid much to the function of insufficient investments in power generation and failure to attract new investments. Faced with this crisis, the government enacted a landmark energy legislation in October 2010 that granted unprecedented powers to expedite energy-related projects [2]. It could be argued that many of the rental power companies initiated at this time for meeting the emergency power crisis fall under this act which raise serious controversy about the transparency of this act as regards to bypassing the competitive bidding process for procuring power from private companies for potential corruption and rapidly raising the cost of power generation. Additionally, the government decided to install more coal power plants from 2014 to phase out costly rental and quick rental power plants with the intention of reducing the power generation costs and the large-scale subsidy paid to quick rental and rental power plants to mitigate any short-term demand crisis. For example, in 2011 about more than $$0.56 billion was paid out the rental power plants due to their high imported cost of fuel.

Table 3.2 provides a snapshot of the progress so far in achieving the strategic objectives of the energy sector during the 6th Five Year Plan or the Results Framework (RF) target. As could be seen here, there has been a substantive progress in terms of expansion of generation capacity over the years.

Figure 3.1 gives a view of the increase in installed capacity including captive power[1] from the period 2015 to 2040. The rationale for this is that the demand for power is constantly increasing in line with the country's accelerated pace of economic growth and fast pace industrialization. It is suggested that Bangladesh aspires

---

[1]A captive power sixth plant is an electricity generation facility used and managed by an industrial or commercial energy user for their own energy consumption.

Table 3.2 Energy sector objectives, performance indicators and targets for the 6th Five Year Plan

Objectives/performance Indicators	FY 2010 (baseline)	FY 2014 (estimate)	FY 2015 (target)
Make Power Sector Financially Viable	Subsidy Tk. 12 billion	Subsidy Tk. 61 billion	Surplus
Increase generation capacity of electricity	5823 MW	10,618	15457 MW
Increase efficiency of energy use as well as reducing system loss	17% T&D loss (FY 08)	14% T&D loss	
Diversifying fuel use in power generation, i.e. from gas to coal, liquid fuel	83% gas & 8% fuel oil (FY 2009)	72% gas & 18% fuel oil	
Increase private sector investments in electricity, gas, and other energy supply	26% of generation capacity (FY 2008)	42% of generation capacity (48% with imported 500 MW)	
Encourage energy trade including energy cooperation with neighbors	0 MW	500 MW	
Finalize the coal policy	Not done	Not done	
Per capita consumption of electricity	170 kWh	270 kWh (FY 2013)	390 kWh
Access to electricity	47%	62% (FY 2013)	65%

Source: Ref. [5]

Demand (incl. Captive Power) + Reserve Margin (MW)

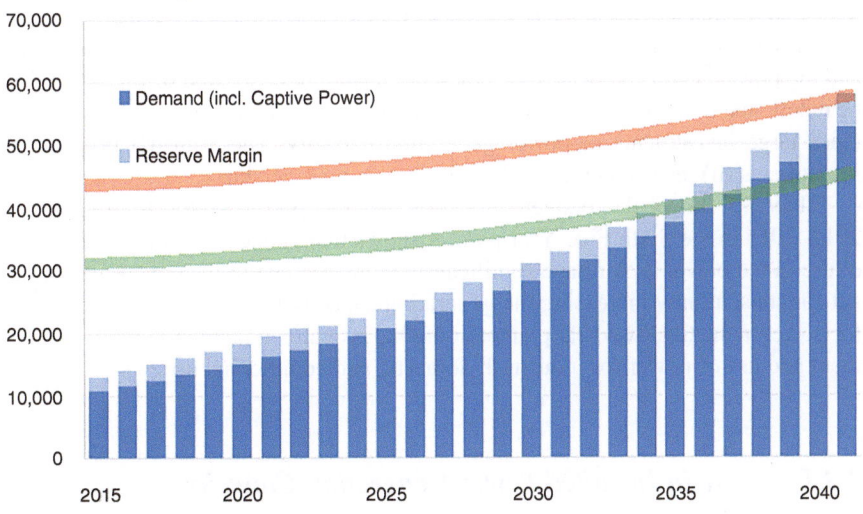

Fig. 3.1 2015 Master Plan peak demand forecast 2040. (Source: Ref. [6])

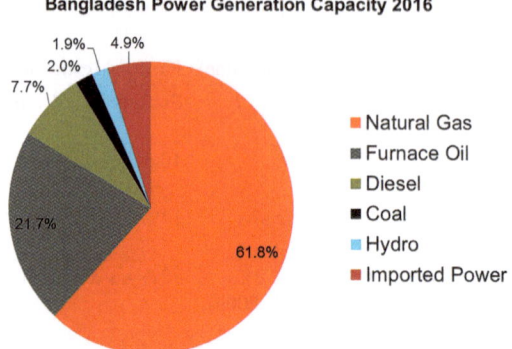

Fig. 3.2 Bangladesh electricity generation capacity transformation 2016–2021. (Source: Ref. [7])

to be a middle-income country by 2021 and a high-income country by 2041. The increase in power generation capacity including an uninterrupted supply of power and energy is a catalyst for achieving its economic development targets.

The Sixth Five-Year Plan projected a fuel diversification which aimed to acquire a fuel composition ratio of 50% (i.e., 30% domestic coal and 20% imported coal), 25% natural gas (including LNG), 5% liquid fuel and 20% nuclear energy including the adoption of renewable energy and cross-border trade. Figure 3.2 summarizes the planned fuel mix of the Sixth Five-Year Plan.

Likewise, significant budget and ADP allocation has been raised in the 6th Five Year Plan to finance the capacity targets as projected in the PMSP. It has been suggested that energy and power investments received most of the budget allocations and the power sector has been very efficient in using the allocated resources. It is to be mentioned that the 6th Five Year Plan targeted two primary sources for financing these large investments in power generation capacity. These are (i) private IPP investments (ii) investment through public-private partnership.

Table 3.3 shows the budget allocation for the power sector for the period 2011–2015 in the Sixth Five-Year plan. It shows that budget allocation and expenditure in the power sector have significantly increased due to the government's target for substantially increasing the generation capacity for meeting the rising need for power in the 6th Five-Year plan period. This is owing to rapid industrial development and use of power for industrial, domestic and commercial use.

3.3.1   Raise in Installed Power Generation Capacity

Following the adoption of PSMP 2010 and its implementation during the sixth Five-Year Plan between June 2010 to June 2014, the total installed capacity has increased from 5823 MW to 10,618 MW. This amounts to an annual growth of 16% per year

Table 3.3   ADB allocation for power and energy sector during the Sixth Five-Year Plan

Taka Billion

Ministry/ Division	FY2011			FY2012			FY2013			FY2014			FY2015		
	Allocation		ADP Expenditure	Allocation		ADP Expenditure	Allocation		ADP Expenditure	Allocation		ADP Expenditure	Allocation		ADP Expenditure
	ADP	SFYP		ADP	SFYP		ADP	SFYP		ADP	SFYP		ADP	SFYP	
Power division	50	50	59.1	71.9	70.7	71.6	85.6	85.6	88.5	79.3	109	78.3	92.7	134.6	83.3
Energy & Mineral Resources Division	11	10.8	9.7	7.3	15.5	7.5	16.1	17.2	16	19	20.1	18.4	22.2	22.9	18.1
Total Energy	61	60.8	68.8	79.2	86.2	79.1	102	103	104.5	98.3	129	96.7	115.0	157.5	101.4

Source: Ref. [5]

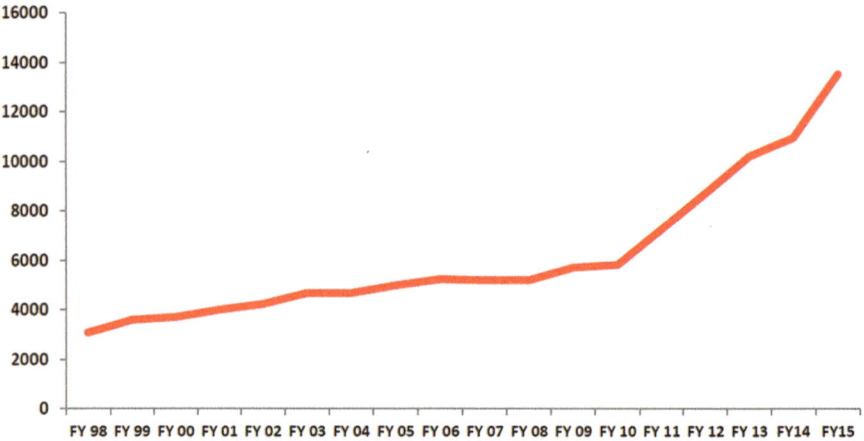

Fig. 3.3 Trend in Grid Based Installed Capacity (MW). (Source: Ref. [5])

as compared to less than 5% achieved in the decade of 1999–2009. Figure 3.3 shows the increase in grid-based installed capacity in MW from 1998 to 2015.

The population access to electricity has increased from the 2010 baseline of 47% to 62% in 2013 and as of September 2015 stands at 75%, which surpasses the 65% target set in the plan. Per capita electricity consumption has also increased from 170 kWh to 270 kWh though this is still considered one of the lowest in the world and lower than most of the South Asian countries [5].

3.3.2 Strategies Adopted in the Sixth Five-Year Plan to Meet the Development Targets

3.3.2.1 Mobilization of Private Investment in Power Generation

Due to constraints in the public sector for funding such large investments, the government aimed to secure a substantial increase in power generation through the private sector in the Sixth Five Year Plan. The new private power sector power generation policy (PSPGB), which had been approved in 2006 formed the principle catalyst in the substantial increase in generation capacity by the private sources. For example, the share of private power supply in terms of installed capacity increased from 26% in 2008 to 42% in 2013. Table 3.4 shows the share of installed power generation capacity in MW in terms of public, private and power imports. It could be seen that the private sector power generation capacity in terms of IPPs have increased considerably in relation to the public sector and playing a dominant role in the combined share of output for power generation. This is due to more favorable investment policies for private power generation conducted by the government in recent decades.

Table 3.4 Installed capacity by ownership excluding captive power (MW)

Year	Public	Private	Import	Total	Private as % of Total
2010	3719	2104		5823	36%
2011	4027	3237		7264	45%
2012	4910	3806		8716	44%
2013	5400	3751		9151	41%
2014	5812	4104	500	10,416	39%
2015	6020	5012	500	11,532	43%

Source: Ref. [5]

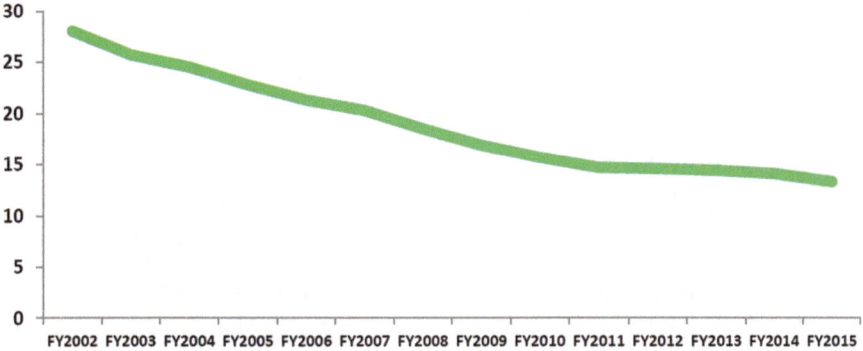

Fig. 3.4 Transmission and Distribution losses in percent. (Source: Ref. [6])

3.3.2.2 Power Sector Efficiency

There has been a remarkable improvement in the reduction of Transmission and distribution (T&D) losses, which is an important indicator of energy efficiency. The T&D loss fell from a high of 32% in 1999–2000 to 17% in 2000. It declined further to 14% in 2014. Along with a large reduction in T&D losses, improvements have been made in reducing power outages, increasing the efficiency of billing and collections and cutting back on accounts receivables. Additionally, the sector's governance and management have been markedly improved [4, 5].

Figure 3.4 gives an overview of the improvement of the T&D losses from 2002 to 2015

Along with a large reduction in T&D losses, improvements have been made in reducing power outages, increasing the efficiency of billing and collections and cutting back on accounts receivables. These have had a beneficial effect on the sector governance and management [5].

3.3.2.3   Power Trade

One of the most remarkable progress in the 6th Five-Year Plan is the import of electricity from India and the initiation of regional power trade. Bangladesh started importing 500 MW power from Bahrampur, India from 2013. This import was initiated by a 400 KV transmission line and HVDC substation from Baharampur and Bharamara of India. During this time several talks were initiated for opening up new avenues for cross border electricity generation with other neighboring countries, especially Nepal, Bhutan and Myanmar. Additionally, the government planned to import 500 MW power from the Rakhain state of Myanmar through the setting up of a hydroelectric power plant in Myanmar on the Lamroriver. The project was supposed to come online between 2015 and 2017. Talks were also initiated on the prospect of importing 1000 MW of electricity from India through a transmission line for which Bangladesh would draw a wheeling charge for electricity to be transmitted through its territory conducted between the two Indian states namely, Assam and Bihar and buy 20–25% of the power from the proceeds. Similarly plans were initiated through two sub-regional initiatives for example, the Bangladesh-India-Bhutan and the Bangladesh-India-Nepal for developing two separate 1000 MW hydroelectric projects by joint initiatives. However, concerns remain about the transmission of power through these initiatives as any power transmitted has to be consented by India sharing borders with these countries and has to match the system development plans of India [5].

3.3.3   Concerns of Achieving the Targets in the Sixth Five-Year Plan

Though the 6th Five-Year Plan registered significant improvements in the generation capacity with additional capacity addition through the private sector and the launch of the power trade, but this additional capacity has been materialized through a substantial rise in the marginal cost of power. As the most of this raise has been effected through the costly oil-based rental power plants importing fuel whose per unit cost of power generation is higher than other conventional sources. Additionally, concerns were raised about the country's status of primary fuel supply due to the rapid growth of the share of oil-power supply (i.e., rising from 8% in 2009 to 25% in 2013). This is due to a severe rationing of gas as the county's principal primary fuel supply will gradually decline from 2017 and will be completely exhausted by 2040 [8]. The Sixth Five-Year Plan emphasized the need for dual fuel for new power generation facilities as the government has stopped supplying gas to new private power projects. Additionally, the plan envisaged the introduction of coal into the energy mix with significant capacity addition by commissioning large planned domestic coal-based power plants. Though Bangladesh has a large reserve for high quality bituminous coal with low ash and sulfur content with good heat value for power generation, coal mining has not proceeded due to a lack of national coal policy. Furthermore, there were issues in extracting the coal with either open pit or underground mining method with the contention of the best policy to decide for its large-scale extraction. It is to be suggested that none of the coal-based power projects identified under the Sixth Five-Year plan came online during this period with only 250 MW of domestic coal capacity was operational in 2017. The capacity addition as forecasted in the PSMP 2010 fell short of the target in the Sixth Five-Year Plan and a dependence on costly oil-based plants continued for failing to make way to come onstream any large baseload gas and coal power plants for meeting the growth targets and demand for power in the interim measure [9].

3.3.4 Progress of Sector Reforms in the Sixth Five-Year Plan

With a significant increase in power generation, the Sixth Five Year Plan saw through some notable reforms. Some of these reforms were (i) the formation of Sustainable and Renewable Energy Development Authority (SREDA) under the Sustainable and Renewable Energy Development Authority Act, 2012; (ii) the rural electrification board act 2013 (iii) interim action plan for improvement of energy efficiency and conservation (iv) clean cook stove country action plan (v) special act for quick procurement in power and energy sector (vi) electricity amendment act 2014 and (v) energy efficiency and conservation rules. One of the most significant developments in this regard is the formation of the SREDA – a nodal agency under the government that is dedicated to accelerate the development of scalable power generation through renewable energy. It is also entrusted with energy conservation including (i) energy management program (ii) energy efficiency labelling and (iii) energy efficiency building program. Other notable reforms were improved operations and maintenance (O&M) practices, the full and proper functioning of BERC, as the organization has been successfully able to perform its agenda on licensing, energy pricing, quality of utility performance including energy efficiency and consumer satisfaction and dispute resolution measures [4].

3.4 Seventh Five-Year Plan

The Seventh Five-Year Plan (2016–2020) aims to continue on the significant progress made in generation capacity including actual production and capacity and increase the population access to electricity. The Government of Bangladesh targeted GDP growth of 7.4% per year between 2016 and 2020 to accede Bangladesh to a middle-income country by 2021 and have universal power for all. It also targeted to mitigate two shortcomings of the Sixth Five-Year Plan which are (i) increasing the cost of electricity production; and (ii) continued operational deficits in the power sector [10]. It has been suggested that progress made in capacity addition in the Sixth Five Year Plan came on the heels of a significant electricity crisis and the government had to take a more pragmatic approach by using more short-term rental power plants run on liquid fuel as an interim measure to mitigate the crisis. However, the Seventh Five-Year Plan put the emphasis on least cost domestic power generation based on low cost long-term base load power plants run on imported coal and also quickly phasing out the existing rental power plants. Since the least cost power generation approach is dependent on the choice of primary fuel and since the main primary source of fuel – gas, is fast depleting, the PSMP 2010 projected that coal would be the substitute for gas in the future. According to the PSMP 2010 it was projected that coal would rise from 3% by the end of 2015 to 23% during the Seventh Five Year Plan and would subsequently be 44% by 2030 as a dominant source of energy. Based on the experiences of the Sixth Five-Year Plan, the government revised the PSMP 2010 at the beginning of the Seventh Five-Year Plan for having a more realistic assessment of the growth targets projected for the Seventh Five-Year Plan taking into the constraints of the lack of a national coal policy, problems in LNG import including its sourcing, investment costs and logistics considerations and supplying power through nuclear fuel as projected in the Seventh Five-Year Plan. As evidences suggests that Bangladesh has planned to commission two nuclear power plants each 1000 MW on a joint venture initiative between Bangladesh and the Russian Federation by 2017 and 2018 respectively. The new PSMP 2016 has been aligned with the government's revised 2010 recommendations by using a balanced combination of LNG, coal and import of power rather than the heavy dependence on coal for power generation in the Sixth Five-Year Plan [5].

Based on the revised estimate of the PSMP 2010, the Seventh Five-Year Plan set a target to increase installed electricity generation capacity to 24,000 MW by 2021 and have universal power for all. It is to be mentioned that based upon this plan, installed capacity has increased to 23,548 MW by 2020 which includes public sector, IPPs, captive power, imports and renewable energy. Electricity coverage has increased to 97% which was 72% in the baseline in 2015. This has resulted in increasing per capita generation of 512 kWh from 2015. Table 3.5 shows the year wise target of generation capacity from 2016 to 2021.

The investment requirements for power generation for the Seventh Five-Year Plan is substantial when the cost of transmission and distribution are added. It is estimated that the total investment for each year during the plan period would amount to 2.3% of GDP. In line with this ambitious target, the ADP allocations for the power sector has been increased to meet the new investment targets which range from Taka 164.9 billion to Taka 190.4 billion at the end of the period.

Table 3.6 presents an overview of power sector financing requirements both from the government and the private sector in the plan period.

Table 3.5 shows the year wise target of generation capacity from 2016 to 2021

FY	Target (MW)	Target for Access to Electricity (%)
2016	14,943	80
2017	16,399	85
2018	19,249	90
2019	20,649	94
2020	23,000	96
2021	24,000	100

Source: Ref. [6]

Table 3.6 Financing requirements in the power sector from 2016 to 2020

						Taka Billion
Ministry of Energy and Mineral Resource Division	FY 2016	FY 2017	FY 2018	FY 2019	FY 2020	Total
	19.9	34.5	41.1	48.2	56.6	200.3
Power division	164.9	168.5	171.5	201.0	235.9	941.8
Total	184.8	203.0	212.6	249.2	292.5	1142.1

Source: Ref. [5]

3.4.1   Strategies for Attaining the Growth Targets in the Seventh Five-Year Plan

To meet the growth targets for the 7th Five-Year Plan and have universal access to power for 2021, the government set a number of strategies for meeting these growth targets. These are discussed below.

3.4.1.1   Mobilizing IPPs

As the plan envisaged that it would require $22 billion in new generation capacity with significant financial constraints in the public sector to mobilize such investment, the government has to increasingly rely on new IPPs including the installation of large base-load power plants to meet future capacity targets. The primary goal of this plan was to move away from the existing rental power plants to low cost base-load power plants on a path toward the least cost option principal for setting up power plants and the concern for supply of primary fuel for meeting such large power generation capacity targets.

3.4.1.2 Electricity Expansion through Renewable Energy

The 7th Five-Year Plan targeted 10% of electricity generation through renewable energy especially wind and solar as part of its fuel diversification scheme for supplying power to all. The Renewable Energy Policy 2008 and the establishment of SREDA gave the impetus to further accelerate the need for the implementation of scalable power through IPPs and access to power to the rural areas where the grid is not available. It is suggested that about 4.1 million SHS have been installed in this period through Infrastructure Development Company Ltd. (IDCOL) – a government non-banking financial institution through the support of multilaterals and government financial institutions. However, there were concerns about the expansion of renewables which could be attributed to limited land availability and the resource assessment of wind power is still evolving [10]. Additionally, there is a need to develop technical rules and regulations for grid-connected renewable energy capacity including smart grid technology which are at a nascent stage of development for wider integration of renewables. Table 3.7 shows the renewable energy potential of Bangladesh. It is suggested that Bangladesh has good potential for renewable energy development. Especially solar, wind and biomass. Bangladesh does not have geothermal potential and its hydro potential especially based on elevation is small. Solar energy is regarded as the single most dependable renewable energy that can be resourced on a large-scale in the form of grid-tied utility scale solar parks and industrial rooftop projects. It is estimated that more than 1000 MW of utility scale solar parks and 500 MW of commercial industrial rooftop solar photo voltaic (PV) projects under net-metering scheme are on the horizon [11].

Table 3.7 shows the renewable energy potential of Bangladesh. It is suggested that Bangladesh has good potential for renewable energy development. Especially solar, wind and biomass. Bangladesh does not have geothermal potential and its

Table 3.7 Renewable energy potential in Bangladesh

Technology	Resource	Capacity (MW)	Annual Generation (GWh)
Solar Park	Solar	1400*	2,000
Solar Rooftop	Solar	635	860
Sole Homes Systems (SHS)	Solar	100	115
Solar Irrigation	Solar	545	735
Wind Park	Wind	637**	1250
Biomass Generation	Rice husk	275	1800
Biogas Generation	Animal waste	10	40
Waste to Energy	Municipal waste	1	6
Small hydro power plants	Hydropower	60	200
Mini-grid, Micro-grid	Hybrid	3***	4
Total		3666	7010

Source: Ref. [12]

* Case 1 (agricultural land excluded) estimate ** Case 1 (flood prone land excluded) estimate *** Based on planned project only, not a theoretical maximum potential because there is potential overlap with off grid solar systems. Either could be used to serve off-grid demand [12]

hydro potential especially based on elevation is small. Solar energy is regarded as the single most dependable renewable energy that can be resourced on a large-scale in the form of grid-tied utility scale solar parks and industrial rooftop projects. It is estimated that more than 1000 MW of utility scale solar parks and 500 MW of commercial industrial rooftop solar photo voltaic (PV) projects under net-metering scheme are on the horizon [11].

To accelerate investment in this sector, evidence shows that presently a large number of local and foreign companies are submitting proposals in solar power projects ranging from 5 MW to 100 MW. It is estimated that $ 2.78 billion is required to implement small and large-scale projects in the country with the funds being mobilized by the multilateral partners and the government and private sector. As of 2016, renewable energy generation installed in the country is estimated to be 430 MW with the 230 MW Kaptai Hydropower project supplies the bulk of this energy generation which is supplemented by off-grid SHS in rural areas amounting to 175 MW, urban rooftop solar power accounting to 15 MW and a small amount of power being provided by biogas and biomass-based captive plants. Though the country has good prospects for installing renewable energy power plants for wind and solar energy for which it receives a sufficient amount of solar radiation daily (4–6.5 kWh/m^2) with the potential to operate concentrated solar power (CSP) plants. But concerns remain about land availability and fluctuating wind speed in different parts of the country with assessment of scalable wind power still undergoing [5]. Table 3.8 shows the future forecasted renewable energy installation capacity targets from 2025 to 2041.

3.4.1.3   Demand Side Management

The seventh Five-Year Plan has laid considerable emphasis on demand side management or conserving power. It has been suggested that a well-articulated demand side management (DSM) policy is a cost-effective way to curtailing peak demand. For example, Bangladesh's electricity peak demand in the summer months reach at over 7000 MW which causes large load shedding. Therefore, the seventh Five-Year Plan envisages conserving energy in the range of 1000 MW by working in concert with its nodal agency SREDA to meet this development target. Further, the program also included certification mechanism of energy manager and energy auditor and designating large energy consumers in the industry and building sectors for obligation to nominate energy managers who would be implementing the energy program.

Table 3.8 Proposed renewable energy installation capacity targets to 2041

	2025	2030	2035	2041
Electricity generation demand (Twh)	144.32	205.57	322.84	446.02
Target of 10% generation demand from renewable energy (TWh)	14.43	20.55	32.28	44.6
Target of total installed capacity from RE (GW)	10.29	14.66	23.03	31.32

Source: Ref. [13]

The program entailed submission of annual energy reports and implement energy audit for their respective establishments periodically [4].

3.4.2   Structure of the Electric Power Sector of Bangladesh and Future Unbundling

The power sector of Bangladesh is primarily regulated by the Bangladesh Power Development Board (BPDB) under the guidance of the Ministry of Power and Mineral Resources (Fig. 3.5).

Here, BPDB acts as a single buyer. The Power grid company of Bangladesh (PGCB), the transmission operator obtains the power from BPDB and collects a wheeling charge for delivering the electricity at a certain predetermined voltage at the distribution companies' main distribution points. These companies include the Dhaka Power Distribution Company (DPDC), Dhaka Electric Supply Company (DESCO), the West Zone Power Development Company (WZDPC) and Rural Electrification Board (REB). Although in theory BPDB acts as a single buyer, however, with the introduction of the new commercial power development policy 2008,

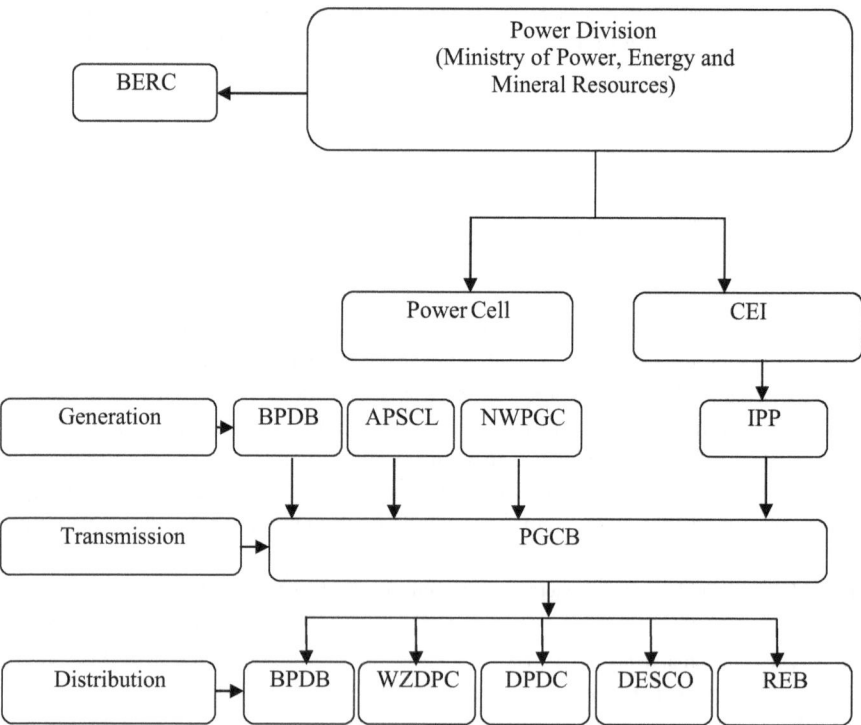

Fig. 3.5 Gives a pictorial overview of the institutional structure of the Bangladesh power sector. (Source: Ref. [14])

private power companies can supply buyer to large consumers on mutually negoti-ated tariffs and also sell 20% of its power to the state companies.

Bangladesh power sector has evolved from a vertically integrated utility to a partially unbundled sector with private entry and competition by IPPs in generation and to a lesser extent in distribution. Over the years the sector has been unbundled both horizontally and vertically with the BPDP acting as a single buyer along-side several generation companies namely, Ashuganj power station company ltd. (APSCL), IPP and Small independent power producer (SIPP), one transmission company and several distribution entities. The BPDB operates about half the power generation capacity while it purchases power from the other three generation com-panies including the private sector as a single off-taker. On the distribution side there are 72 rural cooperatives (Palli Bidyut Samatis) under the BREB that distrib-ute about 40% of the total power, DPDC is 19%, DESCO is 11% and WZDC being 6% and the rest 25 is distributed by BPDB's distribution zones.

Currently the sector unbundling is on hold and it has been suggested that any future unbundling initiatives are met with increased resistance from the collective bargaining agents (CBAs) and other stakeholders of BPDB. Therefore, as an alter-native to further corporatization, the BPDB is more inclined toward implementing Strategic Business Units (SBUs) under its corporate umbrella. Under this initiative each SBU would operate as quasi-independent units with its own board and man-agement structures, separate accounts and a performance management system. These steps have been suggested to improve the operational efficiency of the power sector with reduction in system loss and increased revenue generation [15].

3.4.3 Progress Made in the Seventh Five-Year Plan

The power generation capacity has improved in the Seventh Five-Year Plan from 11,534 MW in FY 2015–2016 to 20,383 in FY 2019–2020. This capacity enhance-ment was led by the private sector in contrast to the public sector led development in the 6th Five Year Plan. At the end of the Seventh Five-Year Plan the population access to electricity stood at 97%. For example, in 2017 the access to electricity had improved to 80% due to the connection of new grids in rural areas by connecting consumers to the network through BREB and its system of rural electrification cooperatives i.e., the PBSs. Though this has been a significant development, how-ever, the universal access to electricity has not been achieved. Two notable acts have been approved in the Seventh Five-Year Plan. These are: (i) electricity act 2018; and (ii) quick enhancement of electricity and energy supply (special provisions) act 2018 which has been amended till 2021 [10].

In addition, cross-border electricity with the India has made a notable progress during this period with 1160 MW electricity has been imported during the plan period. A further discussion had been ongoing to import hydropower from India, Bhutan and Nepal with a trilateral agreement with India to use the Indian transmis-sion system to import electricity from Bangladesh [16]. Further the Transmission and Distribution (T&D) losses have gradually declined from 13.1% in 2015–2016 to 11.23% in 2019–2020. Further Bangladesh started importing LNG from 2018 with the installation of the first Floating Storage Regasification Unit (FSRU) at Moheshkali, Chittagong with a capacity of 3.75 million tonne per annum (mmtpa) importing LNG from Qatar. In that line two new large gas-based combined cycle generation plants with a capacity of 1200 MW each estimated at US$ 3 billion would be run on additional imported LNG. A second FSRU has been installed in April 2019 with Qatar LNG under the agreement of Summit LNG Terminal and Excelerate Energy. A land-based LNG terminal is to be initiated with a 7.5 mmtpa capacity on a private basis constructed by Exxon Mobil, Qatar Petroleum and a consortium of 12 firms with a capacity of 7.5 mmtpa. This has spurred an interest for the government to approve a new policy for importing LNG on a private basis while limiting its sales to the state oil company Petrobangla. Further installation of two land-based LNG terminals are to be initiated between Petrobangla with a Chinese and Indian companies building a 7.5 mmpta LNG terminal onshore at Cox's Bazar – on the south east coast of Bangladesh and agreement has been reached between BPDB and the Saudi private power company AWCA Power for setting up a 3600 MW gas-to-power complex which would house its own LNG import facility [17].

There are several concerns have been revealed at the end of the Seventh Five-Year Plan. One of the principal concerns was over-generation capacity as it has been suggested that over the years the actual generation has been increasingly falling short of the installed capacity which has resulted in large-sized capacity payments to IPPs for not buying their power. This has been significantly increased during the COVID-19 period. Other areas of notable concerns include (i) under-utilization of power plants, which has prompted the off-taker (BPDB) to pay a minimum capacity payment to the IPPs which has been substantial and have been increasing over the years (ii) low levels of efficiency for power plants, for example, an estimated 45% of power plants worked at an efficiency level below 60% due to large variations in plant factors among the power plants (iii) the off-takers continuing increase in yearly expenditure due to rising capacity payments to IPPs, high costs in running oil-based rental power plants and the high cost of importing various fossil-fuels like petroleum, LNG and coal for running power plants [10].

3.5 The Eighth Five-Year Plan and Future Challenges for Achieving the Development Targets

The Sixth and Seventh Five-Year Plan were intended to increase generation capacity to meet demand with special emphasis on meeting the Bangladesh Government's Vision 2021 initiative for having universal power for all and accede to a middle-income country enumerated in the seventh Five-Year Plan. This has resulted in surplus power at the end of the Seventh Five-Year Plan. The Eighth Five-Year Plan aims to follow through on the developments on the Seventh Five-Year Plan and bridging any identified gaps so that these are in line with the goals and strategies laid out in the PSMP 2016 and the future perspective plan (PP 2041) development targets for becoming a develop economy by 2041. It is been estimated that demand for power would grow 8% plus rate during this period. One of the primary objectives of the plan would be to diversify manufacturing and export base with an increase in investment rate from 32% of GDP in FY 2020 to 37% of GDP in 2025. The Eighth Five-Year Plan for enhancing power generation capacity is built on two themes (i) to build a fiscally sound and efficient least cost power generation system with the increase of renewable energy (ii) generating an optimal primary fuel mix using coal (i.e., domestic and imported), LNG and large-scale imported electricity as competitive alternatives in the generation mix. Based on these a revised set of generation targets have been set with 24,000 MW by 2021, 40,000 MW by 2030 and 60,000 MW by 2041 [17].

A revised power generation plan 2030 has been prepared from 2020 and 2030 based on these competitive targets and a new PSMP is set to be in place in line with these new targets. The Eighth Five-Year Plan intends to phase out all the existing rental power plants run on liquid fuel (i.e., furnace oil and diesel) and small-scale IPPs to be replaced by larger and more cost-efficient gas-based and large sized coal fired plants both in the public and private sector. With regard to coal it has been suggested that thought the government has planned to increase coal in the generation mix by 35% by 2041 with the evidence that a large number of coal-fired power plants are different phases of implementation (i.e., the 6000 MW new coal hub in the phase of development in the island of Moheshkhali in the southern part of the country) but evidence suggests that none of coal projects have been completed so far and only 250 MW of domestic coal capacity was operational in 2017. It has been suggested that Bangladesh could face significant challenges when large-scale coal-based power plants come online especially in regard to coal imports notwithstanding the significant capital and logistics costs for importing coal but with the considerable challenges in importing coal including a whole ecosystem in coal use, i.e., transportation and industries and possible locking-in to a high carbon intensity growth path. This could have serious environmental effects for example, air and water pollution and be detrimental in fighting the effects of climate change including applying a social cost of carbon in investment [9]. Additionally, the Eighth Five-Year Plan has put considerable emphasis on developing the local coal base especially consideration in developing coal mining with the respect to some of the challenges in importing coal and its availability for long-term and the adoption of a national coal policy currently not in place for greater use of domestic coal for gaining optimum cost advantage for coal use in power generation in the future. Counterintuitively, it has been argued that since there are significant challenges in installing coal-fired power plants such like (i) suitable land (flood-safe sites for installing large-based load power plants; (ii) the shadow price of carbon for climate change considerations; and (iii) mitigating the effects of global climate change including greenhouse gas emissions (GHG) and pollution - the use of coal in the 8th Five-Year Plan, there is a rethinking of significant use of coal in future power generation for possible cancellation of newly planned projects and delay in the move towards using coal

[9]. For example, the government has cancelled to build ten coal-fired power plants in 2020 and only those plants that are in their various stages of completion are currently being implemented [18]. The other reasons for a reconsideration of setting up future coal-fired power plants are (i) its high environmental impacts; (ii) the reluctance of donors and financial institutions to fund future coal-fired projects; and (iii) and the increasing cost of coals in the international markets.

Additionally, as the power generation in the 8th Five-Year Plan would be relying on imported LNG as an alternative fuel for powering the large installed base of gas-fired power plants due to rapidly declining gas reserves, it has been suggested that supportive infrastructure needs to be set up for LNG import, regasification and storage of LNG including a broader strategy that add to LNG pricing, the optimality of multiple terminals, collaboration and coordination of different agencies for investment in infrastructure, forecasting future demand and flexibility on obtaining LNG on better terms from international LNG markets. With respect to some of the challenges of securing long-term primary fuel supply for power generation in the Eighth Five-Year Plan and beyond and their ready availability and the significantly increasing costs of power (i.e., domestically produced natural gas, coal and imported LNG) cross-border power trade with India has been given a renewed importance for procuring future sources of primary power supply. In this regard the Eighth Five-Year Plan suggests a wholesale power pool market to be developed with India for cross-border electricity imports and going beyond the current Government to Government (G2G) power trade with India. In that regard the development of a joint power pool (i.e., participating in Indian power market or having its own power exchange) with buyers and sellers from both sides trading electricity on a spot price basis would give the impetus of market-based power and greater competition for reducing power costs and system stability for the future [9].

One notable addition to the primary fuel base for the 8th Five-Year Plan would be nuclear power. In this regard it is expected that by the end of the 8th Five-Year Plan the Rooppur Nuclear Power Plant built by the Russian state nuclear corporation Rosatom with its two generation units i.e., Unit 1 with a capacity of producing 1200 MW of power and the Unit 2 with another 1200 MW adding 2400 MW of nuclear power would be coming online in 2023 [17].

3.6 Summary

This chapter presents an overview of the Bangladesh power sector from post-independence starting from 1972 through post 2000 including the Sixth Five-Year Plan (2011–2015) and 7th Five-Year Plan (2016–2020) and beyond. It details some of the reform initiatives during this period including sector unbundling, different policies and acts introduced in these periods (i.e., sixth, seventh and eights five-year plans) the fuel mix and growth targets and some of the challenges and concerns for meeting the capacity targets. It also details some of the adjustments made in government policies in a realistic setting in light of the constraints for attaining the development targets including having universal access to power for all.

References

1. World Bank (2021) Living in the light: the Bangladesh solar home systems story. https://openknowledge.worldbank.org/handle/10986/35311. Accessed 10 Aug 2022
2. Ebinger CK (2011) Energy and security in South Asia: cooperation or conflict? Brookings Institution Press, Washington, DC
3. ADB (2009) Sector assistance program evaluation for Bangladesh energy sector. https://www.adb.org/documents/sector-assistance-program-evaluation-bangladesh-energy-sector. Accessed 15 Sept 2022
4. Alam (2015) Strategy for infrastructure sector: background paper for the seventh five year plan. https://www.resourcedata.org/dataset/rgi-strategy-for-infrastructure-sector-background-paper-for-the-seventh-five-year-plan. Accessed 20 July 2022
5. GOB (2015) Seventh Five Year Plan FY 2016- FY 2020: accelerating growth, empowering citizens. https://www.unicef.org/bangladesh/sites/unicef.org.bangladesh/files/2018-10/7th_FYP_18_02_2016.pdf. Accessed 5 Oct 2022
6. JICA (2016) Survey on power system master plan (Draft final report). http://www.powercell.gov.bd/site/page/b93738b7-c6d2-4bcc-a7cd-298cbb91e6b4. Accessed 15 June 2016
7. IEEFA (2016) Bangladesh electricity transition: a diverse, secure and deflationary way forward. https://policycommons.net/artifacts/2251258/bangladesh-electricity-transition/3009955/. Accessed 5 Aug 2022
8. CPD (2013) Bangladesh economy in FY 2013–14: third interim review of macroeconomic performance. Centre for Policy Dialogue Dhaka Bangladesh
9. Pargal (2017) Lighting the way: achievements, opportunities, and challenges in Bangladesh's power sector. https://esmap.org/node/174806. Accessed 4 Aug 2022
10. GOB (2020) Eighth Five Year Plan July 2020 – June 2025: promoting prosperity and fostering inclusiveness. https://policy.asiapacificenergy.org/sites/default/files/Eighth%20Five%20Year%20Plan%20%28EN%29.pdf. Accessed 7 Sept 2022
11. Daily Star (2023) The future of renewable energy in Bangladesh. https://www.thedailystar.net/recovering-covid-reinventing-our-future/developing-inclusive-and-democratic-bangladesh/news/the-future-renewable-energy-bangladesh-2965606. Accessed 5 Jan 2023
12. SREDA (2015) Scaling up renewable energy in low income countries: investment plan for Bangladesh. https://www.climateinvestmentfunds.org/sites/default/files/meeting-documents/bangladesh_srep_ip_final.pdf. Accessed 7 July 2022
13. USAID (2021) Recommendation for a renewable energy implementation action plan for Bangladesh. https://pdf.usaid.gov/pdf_docs/PA00XD5J.pdf. Accessed 20 Aug 2022
14. Bangladesh power development board. www.bpdb.gov.bd. Accessed 6 Aug 2022
15. World Bank (2015) International development association project appraisal document on a proposed credit in the amount of SDR 155.4 million (US$ 217 million Equivalent) to the People's Republic of Bangladesh for the Ghorashal Unit 4 Repowering Project. http://documents.worldbank.org/curated/en/569211468196182901/pdf/PAD1422-PAD-P128012-IDA-R2015-0302-1-Box393264B-OUO-9.pdf. Accessed 12 Nov 2022
16. CPD (2021) Bangladesh economy in FY 2020-21: interim review of macroeconomic performance. Centre for Policy Dialogue, Dhaka, 1209
17. JR Ichord (2020) Transforming the power sector in developing countries: geopolitics, poverty, and climate change in Bangladesh. https://www.atlanticcouncil.org/in-depth-research-reports/issue-brief/transforming-the-power-sector-in-developing-countries-geopolitics-poverty-and-climate-change-in-bangladesh/. Accessed 5 May 2022
18. 3 (.2023) Bangladesh scraps 10 coal power plants: PM tells COP 26. https://www.thedailystar.net/environment/climate-crisis/climate-action/news/bangladesh-scrapped-10-coal-power-plants-pm-tells-cop26-2211496. Accessed 30 Nov 2022

Chapter 4
Determinants

Abstract This chapter identifies the key factors that are important to conduct FDI in the Bangladesh power sector. These key factors are subsumed under four broad categories of investment prospects namely, regulatory, economic and financial, political and social factors. These factors are based on an empirical study conducted by the author which is the first-of-its kind in the Bangladesh power sector on a sample of 25 FDI power companies consisting of 20 conventional thermal power plants (i.e., gas, liquid fuel, dual fuel and coal) and 5 renewable companies (i.e., wind and solar) from 2015 to 2017 in the Bangladesh power sector. To gain insight on the factors influencing FDI, 30 semi-structured one-one in depth interviews were conducted based on four target groups including private company personnel, government officials, multilateral agencies, and academics. This was then followed by a survey to arrive at the results on the key factors identified as 'very important' to 'extremely important' which are detailed in this section. Please see Appendix.

Keywords Methodology · Land · Economic · Political · Dunning

4.1 Four Areas of FDI Determinants

The four areas of FDI determinants namely, regulatory, economic and financial, political and social are based on Dunning's eclectic paradigm and are extended to include institutions. The basic underpinning behind this model is that to invest in a foreign country a firm needs to have three different types of advantages such as ownership, locational and internalization. For FDI to happen all three advantages must be present simultaneously. Examples of ownership advantage are access to patents, specific entrepreneurial skills, scale economies or superior technology. Ownership advantages make it possible to move between different locations and can thus be transferred to a foreign country. For FDI to take place the ownership advantage also has to be profitable for internalization by the firm rather than the market taking care of transactions such as selling or leasing. Finally, there must also exist

T. Mahbub, *Encouraging Foreign Direct Investment (FDI) in Bangladesh's Power Sector*, SpringerBriefs in Energy,
https://doi.org/10.1007/978-3-031-27990-4_4

some forms of locational advantages specific for the geographical location that would trigger actual investment. Locational advantages are country specific and cannot be transferred to another location such as low input costs, existence of raw material or special tax regimes. The eclectic paradigm focuses on economic efficiency as a key determinant of location choice of MNEs. However, economic efficiency only partially explains the location choice of MNEs, as they also require institutional legitimacy to survive and succeed in a challenging foreign environment. From this perspective it has been argued that the nexus of MNE activity and the institutional environment is an analysis of the ability of institutions to reduce transaction costs associated with FDI in an uncertain foreign environment [1]. Therefore, MNEs are motivated to enhance their legitimacy to become isomorphic to the host country institutional environment for example, making adjustment to personnel, image, branding, government relations and other areas to seek legitimacy in these markets [2]. As such the need to integrate institutional factors in FDI theory can hardly be overemphasized. Noting a lock of institutional content in the eclectic paradigm, Dunning argues that it is important to incorporate institutional factors in an extension of the model. Moreover, aside from traditional market motives that might attract FDI in a developing country, such as low labor costs, physical and human infrastructure, or natural resources, notwithstanding where firms have a choice, the institutional environment also has a decisive role in FDI [3, 4]. This issue seems to be more profoundly linked to the underlying mechanisms behind MNEs decisions regarding cross-border investments, tempered by differences in host countries' institutional setups and their quality in attracting or deterring FDI. Dunning argues that it is important to incorporate institutional factors in an extension of the model, and his subsequent research suggests that institutions affect all three paradigm components [5].

4.2 Methodology

The study employs a mixed-method approach, comprising qualitative analysis from semi-structured interviews and quantitative analysis through questionnaires. A mixed-method approach is better suited to understanding complex problems in cases where there is a lack of long-term data series suitable for econometric analysis as the collection of data on FDI in Bangladesh started on a bi-annual basis in 2004. Additionally, data in the power sector is limited in general. To get trusted results from econometric or modelling analysis, a reasonable number of observations are needed to perform the analysis. The sample size is too small to obtain sensible results from modelling analysis. Moreover, a mixed-method approach is better compared to other mono-methodological studies (i.e., either qualitative or quantitative) as it seeks complementarity, development and expansion in the analysis and interpretation of the data. This study used grounded theory and coding to analyze the qualitative data.

A range of techniques was used for a detailed analysis and interpretation of the interviews. The interviews were transcribed and coded to find concepts, categories and robust themes about FDI. First, the data analysis used open coding to identify the selected categories. Then, these categories were compared and contrasted in order to develop more complex and inclusive categories. Finally, these categories were connected to derive robust themes concerning FDI. Detailed case-based memos were drafted to reflect upon the personal narratives as garnered from the interviews and assess the interviewees' experiences based on their reactions from the interviews. A constant comparison analysis was conducted to rigorously assess the similarities, differences, trends, and patterns in the data. Additionally, this helped to systematically question some pre-established ideas related to the hypotheses and propositions regarding the factors that attract or discourage FDI in the power sector and assess the responses of the interviewees against such assumptions. The validation of the data was achieved through on-going consultations with the interviewees. For example, additional data were collected through follow-ups with the interviewees and further clarification through phone interviews and emails. To negate any cognitive bias against confirmation, an active search for disconfirming evidence was undertaken and a discrepant case analysis was drawn align with the participants' accounts [6].

An empirical study of 25 FDI private power companies was conducted. A list of the FDI power companies – 14 of which are in operation, and 11 under construction – was prepared from the Registrar of Joint Stock Companies and Firms and the Bangladesh Power Development Board (BPDB). The sample comprised 20 conventional power companies and five renewable power companies. Out of the 20 conventional power companies, seven were gas based, five liquid fuel, two dual fuel and six were coal based. Of the conventional power companies, two were small (0–50 MW), eight were medium (50–150 MW), and 10 large (above 150 MW). Of the renewables in the solar power category, two companies were medium (1–25 MW) and two were large (above 25 MW). In the wind category, one company (0–20 MW) was small.

To gain insight into factor influencing FDI, 30 Semi-structured, one-on-one in-depth interviews were conducted during December 2015 – April 2016. Four target representative groups were chosen: private company personnel, government officials, multilateral agencies, and academics. A purposive sampling technique was chosen for data collection.

The questionnaire was composed of three parts on firm information, determinants/barriers related to FDI and personal information. It investigated 46 factors from both literature review and expert interviews. First, FDI factors that encourage or discourage FDI in the questionnaire (both determinants and barriers) were primarily developed through the literature review and 10 preliminary expert interviews conducted from April 2015 to July 2015. This offered a first-hand understanding of the area of investigation and identified any missing gaps between theory and the actual conditions. Second, the questionnaire was pre-tested by six experts from the four stakeholder groups through preliminary reviews and suggestions. Finally, the questionnaire was improved by incorporating feedback from these respondents. The

Table 4.1 Characteristics of Sample FDI Power Plants

	Conventional power plants			20	80
Sample (N)	Renewable			5	20
Ownership	Conventional power	JV		11	44
		Wholly owned		9	36
	Renewable	JV		3	12
		Consortium		2	8
Capacity	Conventional power	0–50 MW		2	8
		50–150 MW		8	32
		>150 MW		10	40
	Renewable	Solar	5 KW-1 MW	0	0
			1–25 MW	2	8
			>25 MW	2	8
		Wind	0–20 MW	1	4
			20–100 MW	0	0
			>100 MW	0	0
Fuel	Conventional power	Gas		7	28
		Liquid fuel		5	16
		Dual fuel		2	12
		Coal		6	24
	Renewables	Wind		1	4
		Solar		4	16
Contract period	0–5 years			7	28
	5–10 years			0	0
	10–15 years			3	12
	15–20 years			5	20
	20–25 years			10	40
	>25 years			0	0
Position of respondent	Power companies	Chairperson		1	1.6
		Managing director		3	4.8
		Director		20	32.3
		Manager		8	12.9
	Government	Secretary		4	6.5
		Director		15	24.2
	Multilaterals	Senior energy experts		4	6.4
	Academics			7	11.2

respondents were asked to fill out a five-point Likert scale structured questionnaire [7] ranging from 1—'not at all important'—to 5—'extremely important'.

Table 4.1 shows the characteristics of the power companies. Among the sampled FDI power companies, 80% conduct conventional thermal power generation, and 20% conduct renewable energy generation. Among those in conventional power, the majority, in terms of ownership, are joint ventures (44%) (JVs), while 36% are wholly owned.

In terms of the capacity of the conventional power plants, 40%, 32%, and 8% are large,[1] medium, and small, respectively; the latter fall into the category of the rental power group. Among renewables, solar power is equally divided between large- and medium-sized firms, each representing an 8% capacity. Regarding wind power, there is one small FDI company that has signed a contract with BPDB and is under construction.[2] In the fuel mix composition 7 were gas based, 5 liquid fuel, 2 duel fuel, and 6 were coal based. Among the renewables, one was wind and four were solar power-based companies.

The interviewees for FDI power companies included one chairman, three managing directors, 20 directors, and eight managers. From the government side, there were four secretaries and 15 directors. From the multilateral organizations, there were four energy experts, and from the educational and research organizations, there were four professors. Notably, most of the interviewees (56%) are directors involved in policymaking, research, and strategic investment decisions.

For data analysis, descriptive statistics are used to compute the average scores and standard deviations. To facilitate the analysis, the means are interpreted as follows: (i) not at all important = 1.00–1.79, (ii) slightly important = 1.80–2.59, (iii) fairly important = 2.60–3.39, (iv) very important = 3.40–4.19, and (v) extremely important = 4.20–5.00.[3]

4.3 General Results

The results of the questionnaire analysis show that MNEs do not give equal weight to the four areas of FDI determinants. Figure 4.1 presents the results of the survey.

Here the regulatory dimension is the most influential, followed by economic, political and social dimension. Within the regulatory area, government commitment to contracts is the most important factor (rating score: 4.52), followed by land acquisition/rent/lease and tax exemption. For the economic area, economic growth and development, gas transmission line, and skilled labor scored the highest. For political considerations, the coordination and collaboration of ministries assumes the highest weight, with a score of 4.11, followed by accountability of public officials (4.06) and control of corruption (4.06). Finally, concerning the social area, citizen security and accountability is rated as very important, with a rating of 4.

---

[1] For conventional thermal power plants in the Bangladeshi power sector, capacity ranges are as follows: 0–50 MW (small): 50–150 MW (medium); and greater than 150 MW (large). In solar power, capacity ranges are 5 KW -1 MW (small); 1–25 MW (medium), and greater than 25 MW (large). In wind power, capacity ranges are 0–20 MW (small), 20–100 MW (medium), and greater than 100 MW (large). (Source: selected interviews and correspondence with personnel from the government and private sector).

[2] The project is scheduled for operation in December 2023. Currently about 40% of the project work has been completed. When operational this project will sell power to the Bangladesh Power Development Board for 18 years. The project faced several challenges including change of ownership, mapping of wind speed and securing funds.

[3] Range of means is calculated from standard deviations.

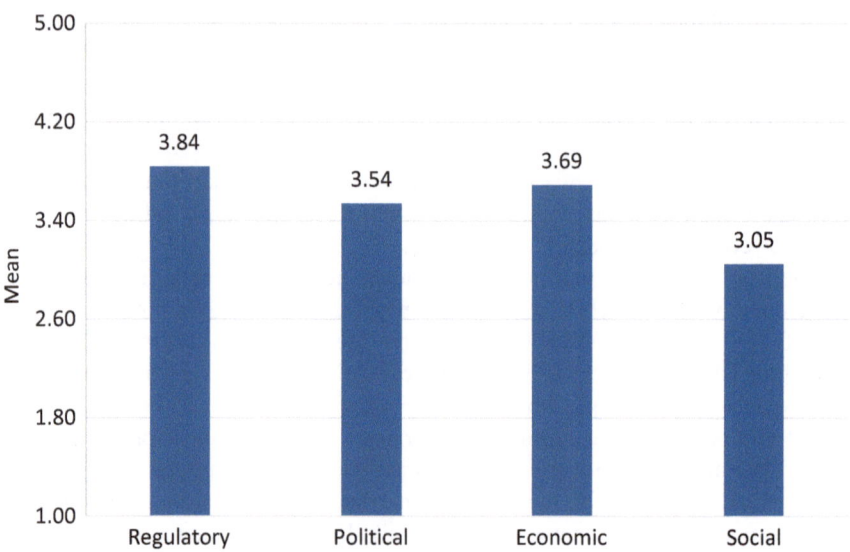

Fig. 4.1 Mean scores of factors categorized by area. (Source: Author's estimates)

These factors are described in detail in Table 4.2. For data analysis, descriptive statistics are used to compute the average scores and standard deviations. To facilitate the analysis, the means are interpreted as follows: (i) not at all important = 1.00–1.79, (ii) slightly important = 1.80–2.59, (iii) fairly important = 2.60–3.39, (iv) very important = 3.40–4.19, and (v) extremely important = 4.20–5.00. In the ensuring discussion only the factors that correspond to 'very important' and 'extremely important' are reported (Table 4.2).

4.4   Regulatory

4.4.1   Government Commitment to Contracts

This factor is rated as the most influential in attracting FDI to the Bangladesh power sector. Foreign investors would like to see that the 'rules of the game' as remaining credible and not altered at the government's convenience once they have made investment decisions based on such rules. This is particularly true in the case of the government's honoring of contracts and regulatory rules [8–10].

However, there are instances when contracts have not been honored. This was seen most vividly in the Asian financial crisis of 1997, where a large number of independent power producer (IPP) contracts were renegotiated either through mutual or unilateral negotiations in countries like Thailand, the Philippines, Indonesia, and Pakistan [11–13]. In addition, when the government makes a pledge regarding the stability of policies and rules, there is sometimes a tendency to behave opportunistically ex post, after the capital has already been deployed [14, 15].

Table 4.2 Mean scores and P-Values of individual factors

Class	Variable label	Mean	*P-value	Class	Variable label	Mean	P-value
Regulatory	Government's commitment to contracts	4.52	0.000	Regulatory (cont.)	Foreign investors can buy shares locally/acquire a local company	3.29	0.002
	Land acquisition/rent/lease of land	4.29	0.000		Regulation on labor	3.18	0.000
	Tax exemption	4.24	0.000		Price cap regulation	3.15	0.001
	Avoidance of double taxation	4.19	0.000		Regulation on subsidy for consumers	2.40	0.000
	Protection of Foreign Investors Act (1980)	4.18	0.000		Regulation on trade union	2.16	0.000
	Profit repatriation controls	4.16	0.000	Economic (infrastructure and financial)	Economic growth and development	4.15	0.000
	Presence of government guarantee	4.15	0.000		Gas transmission line	4.08	0.001
	Time and efficiency of staff to complete the procedure	4.15	0.000		Skilled labor	4.06	0.000
	World class security package	4.13	0.000		Government spending for infrastructure	3.95	0.003
	Protection of property rights	4.10	0.000		Financial facilities	3.94	0.001
	Quick allocation of work permits	4.08	0.000		Credit facilities	3.87	0.042
	Approval of central bank for transferring capital	4.08	0.000		Road network	1.77	0.000

(continued)

Table 4.2 (continued)

Class	Variable label	Mean	*P-value	Class	Variable label	Mean	P-value
	Environmental regulations	4.06	0.000	Political	Coordination and collaboration between ministries	4.11	0.000
	Need for internationally accepted environmental and social impact assessment (ESIA) for large projects	4.06	0.000		Accountability of public officials	4.06	0.000
	Property registration	4.03	0.001		Control of corruption	4.06	0.000
	Regulation on health, hygiene, and safety of workers	4.00	0.001		Capacity to adapt policies	3.97	0.002
	Responsiveness of needs and timeframe of investors	3.98	0.001		Excessive bureaucracy and red tape	3.26	0.017
	Level of administrative competence	3.97	0.002		Electoral process	3.06	0.000
	Construction permit	3.92	0.011		Freedom of press	3.06	0.000
	Continuity and consistency of rules and processes	3.92	0.002		Vested groups	2.77	0.000
	Bangladesh Arbitration Act 2001	3.90	0.022	Social	Citizen security and accountability	3.90	0.000
	Regulation on qualification of personnel who supervise construction	3.87	0.003		Strengthening links between citizens and the government	3.35	0.023
	Regulation on ownership (wholly owned subsidiary/ joint venture)	3.39	0.036		Male predominance	1.90	0.012

Note: Only significant factors are reported in this table
*Note: A one-tailed hypothesis test was performed, and to interpret the results, the P-value was compared to the significance level, which in this case was P-value ≤ 0.05

According to the field interviews conducted by the author it was found that this factor forms the guiding rule seeing their contractual obligations honoured, securing revenues, and creating a level of confidence that investors' interests will be protected. This is illustrated most vividly by the Asian Financial Crisis in 1997 where a large number of IPP contracts were renegotiated either through mutual or cooperative negotiation or unilateral negotiation in countries like Thailand, the Philippines, Indonesia and Pakistan [11–13]. It was found that the country has set a good track record in its experience with IPPs and there have not been any instances of payment default *per se* by the off-taker – BPDB.

4.4.2 Land Acquisition/Rent/Lease

This is the second most important factor in the regulatory category. Access to land is a major issue for setting up power plants in Bangladesh. Foreign investors typically depend on the government to select suitable lands for them, as there is a scarcity of land as well as the complexities involved in the land acquisition process. Access to land and land registration are regulated by multiplicity of laws and regulations that create a maze of rules that are both complex and difficult to administer for foreign investors to apprehend. Moreover, there are other sensitive issues like resettlement and rehabilitation of the affected people who primarily depend on the agricultural land for their livelihood and costly litigation and ownership issues related to land acquisition, as there are multiple titles of ownership for the same piece of land, which makes the entire acquisition process costly, time consuming and risky [16, 17].

Evidence suggests that the cost, availability and difficulty in procuring land in the country are some of the major concerns when making investment choices for setting up power plants in Bangladesh. Additionally, even when the foreign investors are successful in acquiring large tracts of land in a suitable place for setting up power plants, there are several pockets which the landowners do not like to give up and use this as a bargaining tool for raising prices thus creating additional costs for the investors when acquiring land. However, for some investors this had not been a major issue, as they have been provided with lease land from the government as to their strong affiliations with the political process or had the recourse to make available their own private land to complement the land acquisition process, as happened for setting up renewable energy plants (i.e., solar) in the country [18].

4.4.3 Tax Exemptions

This factor is considered one of the most important factors that encourage FDI in the power sector by foreign investors in Bangladesh. The rationale for this this stem from the belief that power companies are exempt from corporate tax for a period of

15 years. Apart from the conventional thermal power plants, renewable power companies (i.e., wind and solar) are also exempt from corporate income tax for 10 years. These measures are drawn as part of the fiscal incentives given by the government to draw large-scale private investments (i.e., IPP) in this sector. As in the future government plans to incorporate coal into the primary fuel mix for generating least cost power replacing the currently running liquid fuel plants with their high average cost of power generation from importing fuel, the government has introduced incentives for coal -based power plants to come under the provision of 15-years tax break if these plants could be brought onstream by June 2020.

Evidence suggests that private companies regard tax exemption as a key factor that is helping to attract FDI. Moreover, this incentive is helping the government to ensure a lower tariff from the private companies which helps for the subsequent mitigation of losses for the off-taker when selling subsidized power to end users [19].

Additionally, FDI power companies are exempt from importing equipment or machinery to set up their power plants. These help for increasing the after-tax cash flow and earnings of companies operating in the power sector including the renewable power companies. Currently for renewable power companies the income tax exemption has bene extended until 2024.

4.4.4 Avoidance of Double Taxation

Bangladesh allows avoidance of double taxation for foreign investors investing in the power sector on the basis of bilateral agreements. It also provides unilateral relief of taxes paid abroad on foreign sources of income [17]. The government has signed double taxation avoidance agreement (DTAA) with 36 countries [20]. It has been argued that the bilateral treaties with different participating countries are not identical as some treaties allow for complete avoidance of double taxation while others provide provisions of reduced taxes if not a complete elimination of taxes. Evidence suggests that not all foreign companies are enjoying this incentive due to a lack of such treaties with their home countries. Additionally, it has been observed that the procedure itself is time-consuming, and often times the National Board of Revenue (NBR) which administers this provision for foreign power companies do not instantly recognize and act on it.

4.4.5 Protection of Foreign Investors Act 1980

This factor assumes high importance as to conduct FDI in Bangladesh. Under the Protection of Foreign Investors Act 1980, Bangladesh is committed to providing non-discriminatory treatment to foreign investors, and FDI companies are to enjoy full protection and security in the country. The act also inclusively assures the full repatriation of capital by FDI entities [17].

Evidence suggests that foreign investors categorically agrees to the principles of this act, which lay down the legal foundations for foreign investment in Bangladesh. However, counterintuitively it has been suggested that the act is very vague and does not clearly define the rules of engagement for foreign multinationals in the local context. This could be attributed to the fact that at the working level, from the experience of foreign investors, it has been found that the country is not as open to FDI as the act broadly suggests. For example, though the country is seen as one of the most open economies in the region with very few restrictions on FDI (i.e., a study conducted by World Bank in 2012 found that out of the 32 sectors for which data were collected, only the forestry sector had restrictions on foreign ownership), however, when conducting business on the ground foreign investors are subject to complex regulatory issues related to entry and establishment, taxation, access to skills and land, foreign exchange regulations, corruption and public governance. These are further compounded by lack of transparency and clarity in regulations and procedures by way of a complex institutional setup [19].

4.4.6 Profit Repatriation Controls

Though Bangladesh allows full repatriation of profits for foreign investors to attract foreign investment, as widely promulgated in its investment promotion policies, in reality, it exerts a relatively strict control over foreign exchange transactions [19]. It has been argued that capital controls prevent investment abroad from Bangladeshi entrepreneurs and restrictions on the convertibility of Taka for business purposes limit beneficial trade and investment in the country [21].

Evidence suggests that foreign investors could fully remit their profits outside the country, although there remained certain concerns. It was revealed that there are additional procedural delays and costs each time the company transfers profits outside. For example, the central bank levies a 5% tax deduction at the source for each transfer of profits. Also, it is compounded by additional bureaucratic delays by the off-taker and the central bank through lengthy verification processes and paperwork, which do not bode well for foreign investors when remitting profit abroad.

4.4.7 Presence of Government Guarantees

In Bangladesh, sovereign guarantees were allowed for IPP projects in the initial launch of the program in 1996 to increase the much-needed capacity and attract foreign capital. It has been argued that in subsequent stages sovereign guarantees have been given to a number of IPPs, especially those facing serious challenges in importing machinery and other related logistics to set up the power plants. However, such moves were widely criticized due to the allocation of such guarantees to companies who supposedly lacked the experience to implement the expected power projects [22]. It is assumed that sovereign guarantees are given by governments to cover a wide range of risks, which are legal, political, regulatory and financial risks. However, a blanket sovereign guarantee that covers all forms of project risks is not feasible and the actual requirement of government guarantee depends on the nature of the project and the extent of the risks involved [23].

Evidence suggests that presently, the government has stopped awarding such guarantees to private IPPs as the off-taker has already established a track record in dealing with local and foreign IPPs. However, it was revealed that many foreign investors seek such guarantees, as lenders consider this when releasing large funds. This is primarily due to the perceived risks of payment default by the off-taker; otherwise, the interest premium on the loan is set at a high level.

4.4.8   Time and Efficiency of Staff to Complete Procedures

This factor is rated as very important in the consideration of making investment decisions in the power sector. Normally, the construction of a power project goes through different stages before the commercial operations start. In Bangladesh, the typical procedure from the achievement of financial closure to the start of commercial operations is about 1–3 years for repeated projects and 2–4 years for large complex projects.

It was identified that once the power purchase agreement (PPA) is signed, the project has to pass through certain formal procedures (i.e., engineering, procurement and construction, lending, O&M, shareholders and fuel supply agreements) of which securing a timely financial close and meeting the official commercial operations date (COD) are extremely important. However, many foreign investors acknowledge their concerns with respect to securing financial close on time, as they had to secure loans both from local and foreign sources, and the expediency and efficiency of their workers to complete the construction on time.

4.4.9   World-Class Security Package

IPPs are governed by a set of contractual agreements, namely, the Implementation Agreement (IA), PPA and the Fuel Supply Agreement (FSA), which form the fundamental security package aimed at reducing lenders and investors' risk, especially when investing in developing countries. Moreover, these agreements form the basis for creating the requisite investment environment and setting the procedures for orderly private sector development in developing countries, especially in infrastructure sectors such as power [24].

It was revealed that it is the security package, developed under the norms of international best practices, that is giving private investors the confidence to show their willingness to participate in the Bangladesh's power sector. It is very liberal with the provision of escrow accounts, land conveyance agreement, O&M and other formal procedures between the parties including liability insurance and international dispute settlement mechanisms such as recourse to International Centre for Settlement Mechanisms (ICSID) and the International Chamber of Commerce (ICC) provisions for FDI attractiveness in the power sector.

4.4.10 Protection of Property Rights

Bangladesh is on a track to graduate from least developed country (LDC) status in 2026, therefore, it is temporarily exempted from obligations from the WTO for strict enforcement of property rights. Bangladesh is a signatory of some of the world's leading intellectual property conventions i.e., WIPO and the Paris Convention of Intellectual Property. However, systematic violations of intellectual property rights have occurred for foreign firms operating in the country in areas such as fast-moving consumer goods manufacturers, film studios, pharmaceuticals, apparel and software firms. It is suggested that intellectual property rights are enforced poorly which magnifies the risk perception of foreign investors. This is attributed to not only the gap in the regulatory framework but to weak institutions and a lack of commitment for enforcement as well. Though just recently a draft Bangladesh Industry-Designs Bill has been approved by the Cabinet to replace the Patents and Designs Act 1911 for making the requisite updates to existing regulations and improving the intellectual property right (IPR) system in Bangladesh, however, it this has not been effected into a law yet. A National intellectual property (IP) policy was developed in 2018 but has not been fully implemented yet. The government has enacted a copyright law in 2000 (amended in 2005), a Trademark Act in 2009, and a Geographical Indications of Goods (Registration and Protection Act) in 2013 including a recent action on bills replacing the Patents and Designs Act [25].

In the field interviews conducted by the author it was revealed that the protection of property rights ranks very high in the decision-making process to conduct FDI in developing countries. Foreign investors generally have a positive attitude toward property rights protection in Bangladesh. It was suggested that the technologies used in the power sector are widely available and not patent protected. Therefore, any kind of infringement, espionage, or leakage is not an issue. However, drawing from the responses it was revealed that protection of property rights could be a significant issue for foreign IPPs in the future as some companies are facing strong competition from other big players in the market, especially the protection of O&M practices which they seek to optimize and protect in order to continue to play a leading role in the industry.

4.4.11   Quick Allocation of Work Permits

This factor is very important for foreign companies that wish to bring in home country professionals and staff to set up and operate their business affiliates. The issue of work permits foreign skilled employees is specifically important to foreign companies that wish to bring in their home country staff such as domestic directors, managers, and specialist for their foreign operations. However, the process of obtaining temporary work permits (TWP) for foreign professionals is relatively long in Bangladesh, officially taking 7–9 weeks with no fast track procedure available [19].

Evidence suggests that it takes on average a month in obtaining work permits. Currently Bangladesh Investment Development Authority (BIDA) which if the successor of Bangladesh Board of Investment (BOI) facilitates the process of issuance of work permits. Additionally, it was also reported that some foreign investors had to resort to paying bribes or other unfair means for faster issuance of work permits.

4.4.12   Approval of Central Bank for Transferring Capital

Bangladesh maintains significant controls over its foreign exchange regime. A 2011 meeting of the International Chamber of Commerce Bangladesh (ICCB) identified that capital controls preventing investment abroad by local entrepreneurs and restrictions on the convertibility of the Bangladeshi Taka are some of the significant rate limiting steps needed to promote doing business in the country. Also, a significant scrutiny and approval process is involved in most foreign exchange transactions [19].

This factor is rated very important in considering invest decisions in the power sector. Evidence suggests that the approval process for transferring dividends abroad is longer and subject to additional documentation requirements and verification. Moreover, there is a cap on the amount to be transferred, and the transfer has to be made consistently from the same bank. In addition, a 9% advance tax liability payment is required with each transfer.

4.4.13   Environmental Regulations

This factor is rated very important for affecting decisions in the Bangladesh power sector. Bangladesh Environment Conservation Act 1995 mandates that all industrial projects must obtain an environmental clearance certificate (ECC) issued by the Department of Environment. Industrial projects are categorized into four groups: green, Orange-A, Orange-B and red in the order of potential environmental risks [26]. Projects that fall into the red category have significant adverse environmental impacts such as fossil fuel-based companies while green having virtually no such issues like renewable power companies. All power projects, therefore, requires among others an Environment Impact Assessment (EIA) for obtaining environmental clearance from the Department of Environment (DOE). This involves three steps. First obtaining the site clearance for approving pre-construction and construction activities, second obtaining approval of the EIA study and third, obtaining the ECC [27]. However, no site clearance is needed for the issuance of the ECC for green projects.

All foreign investors agreed in principle that they have to meet the environmental regulatory standards relevant to the kind of fuel they use in their plants. This issue is currently more stringent for the liquid fuel power companies running on heavy fuel oil (HFO) or diesel which are subject to frequent environmental audits which occur quarterly for the renewal of their environment clearance certificate (ECC). Moreover, there are additional regulations on the import, handing, and storing of liquid fuel at the plant site. One respondent commented that their plant is situated on the bank of a river and carrying the liquid fuel from the port to the plant site is a matter of additional concern for any pillage, mishandling, or leakage with potential grave consequences. This issue is not as much of a concern for other fuel-based plants, namely gas, which have fewer emissions, or some renewable power facilities under construction. However, it does prove a serious challenge for other private, coal-based power plants, which are supposed to come on line in the future, as they have large environmental impacts.

4.4.14 Need for Internationally Accepted Environment and Social Impact Assessment for Large Projects

Foreign investors rate this factor as very important, especially for investing in large FDI power projects in Bangladesh. An environment and social impact assessment (ESIA) are required for projects in which extensive onshore development is proposed with possible long-term impacts on the environment and the people. The framework sets out the guideline for mitigation, monitoring and institutional measures during the design, implementation and operation of the project to eliminate adverse environmental and social impacts, offset them or reduce them to acceptable levels [28]. In addition, it is strictly required by multilateral organizations such as the World Bank and the IFC [29].

The respondent companies that performed ESIA for their power projects revealed that they used a combination of local and foreign consultants in preparing the report, conducted in different phases, with extensive field surveys and modeling work primarily managed by the local consultants in compliance with the local environmental regulations. The foreign consultants mainly undertook the technical review of the report. They maintained that there are good consulting houses within the country capable of preparing such reports according to international benchmarks.

4.4.15 Property Registration

The property registration process in Bangladesh is very slow. According to World Bank's Doing Business Report 2019, Bangladesh ranks 183 out of 190 economies in registering property, and typically, it takes 270 days compared with 69 days in India, 27 days in Indonesia, 53 days in Vietnam, and 9 days in Thailand [30].

Evidence suggests that it takes on average a year to register property in Bangladesh, especially for acquiring newly purchased land from private sources. However, for those companies leasing land from the government for setting up their power plants, the registration time was shorter, taking approximately 1 month on average.

Drawing from the field interviews, the author found that that having a lease from the government is the best option for registering property, but most often these are allocated to companies who nurture strong political ties with the government.

4.4.16 Regulations on Health, Hygiene, and Safety of Workers

The Bangladesh Labor Code 2006 delineates detailed guidelines on the minimum standards for health, hygiene, and workers' safety to be followed on plant premises. This entails requirements for cleanliness, proper ventilation and temperature conditions, exposure to dust and fumes, disposal of waste, lighting, drinking water, and safety issues [31].

It was revealed that proper health, hygiene, and safety conditions are not strictly followed in most of the private power companies due to the lack of awareness of such issues and a tendency for companies to cut supplementary costs. However, countervailing evidence also suggests that western MNEs and JVs give consideration to these regulations by either enforcing them directly as part of their global compliance requirements or at least having them adequately ensured through their local counterparts.

4.4.17 Responsiveness of Needs and Time Frame of Investors

Foreign investors would like to work within a predictable timeframe, especially when bidding for tenders for new IPP projects and would like to see the projects are awarded meeting the best objective criteria (i.e., technical, financial and commercial) within the timeframe of the bid validity and that there are no protracted delays by the government in the award of projects. Any protracted delays could have considerable impact on the overall project cost such as any extension of the bid security which is costly including other concomitant factors for project implementation such as the price of machinery, the financing rate set by the lenders and other costs which could have a direct impact on the final bid tariff. Therefore, any delays in awarding contracts for setting up power plants could increase the final price of projects and a subsequent increase in tariffs which may not hold conducted in the initial negotiations with the off-taker resulting in losing the bid.

4.4.18 Level of Administrative Competence

The level of administrative competence is an important determinant in conducting FDI in Bangladesh's power sector. It has been generally regarded that a cumbersome regulatory environment, onerous procedures and an inefficient public administration are some of the greater stumbling blocks to investment in Bangladesh. This is further supported by a World Bank study [32] where foreign investors cite inefficient government bureaucracy as the third most problematic factor in doing business in Bangladesh. These impose costs on doing business and hurt the country's chances to attract private investment and also compete in global markets.

During the field interviews foreign investors showed their concerns that dealing with customs officials is a perennial problem that they have to encounter more frequently. This cause delays in releasing consignments of necessary spare parts and components for plant machinery having a detrimental effect on plant operation. Moreover, items that are regarded duty free as per the terms of the contract are sometimes subject to additional inspection and scrutiny by the customs authorities, which sometimes entail them to bring out additional custom order requests for those selected items further adding to the delays and costs incurred by the power companies.

4.4.19 Construction Permit

Globally Bangladesh stands at 135 in the ranking of 190 economies on the ease of dealing with construction permits in comparison to 112 for Pakistan, 66 for Sri Lanka and 27 for India [30]. The majority of respondents confirmed that it takes considerable time for approving licenses and permits from the respective government agencies. In addition, a large number of permits needs to be obtained for the construction of power plants, which ranges from 40 to 60 and there is no centralized agency, which could deal with the permits and approvals. Investors have to expend considerable time and also need to pay rents to the government officials for securing the permits. Normally the whole process of getting the permits takes on the average 4–6 months and needs Ministry's recommendations for obtaining the permits. This has significant bearing on the project construction and implementation phase, which need the required permits and approvals for testing and commissioning of the facility within the specified timeframe for completing the project.

However, counterintuitively it was suggested that it did not take them much time for obtaining the licenses and permits as the off taker intervened to having them available for them within a reasonably short period of time. Foreign investors opined that there should be a one-stop service for issuing all licenses and permits, which could fast track the approval process, reduce barriers to timely investment and also put a check on corruption in the system.

4.4.20   Continuity and Consistency of Rules and Processes

Foreign investors would like to see a sense of stability and the continuity of the rules that govern IPPs in the power sector. They would like to see the basic rules and principles are maintained that create confidence to invest in the power sector and there is no abrupt change of policies that affects future sustainability of projects.

For example, there is a general belief that many contracts especially awarded through direct negotiation not through competing bidding may not be honored with any subsequent change of government as previously there was no such policies in the power sector. Another consideration is that the government recently allowed private companies to import liquid fuel on their own instead of procuring from the local state oil company Bangladesh Petroleum Corporation (BPC) with a change of the existing fuel import policy. On that basis, it was revealed that one FDI power company had already invested in building large storage facilities for importing fuel for its two private power plants. However, there has been again a change of policy and the government has stopped private power companies to import fuel on their own as the government had found significant irregularities by a few private companies for importing fuel. It is therefore assumed that such abrupt change of polices and rules have significant impact on investors' costs and profitability and the future sustainability of projects.

4.4.21   Bangladesh Arbitration Act 2001

In keeping with the global norms of international commercial arbitration Bangladesh enacted The Arbitration Act 2001 based on the UNCITRAL Model Law on International Commercial Arbitration (1985) consolidating laws applying to both domestic and international commercial arbitration. The new act creates a single and unified legal regime for arbitration in Bangladesh in line with the recent trends seen elsewhere for commercial arbitration, for example, Germany and India. Though it has certain flexibilities and, in some instances, provides a wider scope and greater powers than the Model Law and the New York Convention in settling commercial disputes, but on the whole its enforcement of foreign arbitral awards is much limited [33].

In the field reviews it was found that foreign investors identified with this act and took it as a positive step in matters of settling alternative disputes related to FDI and it acts as part of the function of settling such disputes in Bangladesh. However, most foreign investors have the recourse to direct international arbitration through International Centre for Settlement of Investment Disputes (ICSID) and International Chamber of Commerce (ICC) provisions built into their contracts for resolving contract disputes. It was found that the later measure is the predominant option if such eventualities do arise taken into consideration the complexity of the processes and the chances of receiving an equitable reward which is more secured in the later options.

4.4.22 *Regulation on Qualification of Personnel Who Supervise Construction*

This factor is rated high in conducting FDI in the Bangladesh's power sector. Evidence suggests that there is a shortage of supervisory and management skills for overseeing construction work for power projects. As supervising construction of large power projects needs sophisticated technical and management skills owing to the large and complex nature of such projects and a higher level of planning, control and coordination dealing with a multitude of parties and workers coupled with time constraints for project completion. It is generally believed that there is a lack of such adequately trained workforce supervising construction of large power projects in Bangladesh. Moreover, IPPs in Bangladesh are managed by foreign Engineering Procurement and Construction (EPC) contractors who engage a large foreign workforce in the construction period of the project. They also help train the local workforce so that after the completion of the project, it is primarily maintained by the local workforce. This is helping the development of local EPC contractors in Bangladesh for managing local power projects and extend internationally.

4.5 Economic and Financial

4.5.1 *Economic Growth and Development*

This factor in the most influential factor in the economic category. Bangladesh has experienced a robust and resilient economic growth performance over the past decade, with real GDP growing at a healthy rate of around 6.5%. Its economy grew by 7.24% in 2020–2021 up from 7.11% in 2016 and recorded a GDP growth of 8.3% in 2019 [34]. This growth was achieved through the accumulation of physical capital and increase in the size of the labor force underpinned by sound macroeconomic management, targeted trade policy reforms, financial sector liberalization and in investment in human capital and protection. Its export grew more than 34% to US$ 52.08 billion in the FY 2021–2022. In addition, the country's current sovereign credit rating of "Ba3 stable outlook" is positive for allowing foreign investors to invest in the country as a whole. In addition, this improved credit rating is bringing the cost of borrowing (i.e., interest) down from other external sources of financing like ECA (export credit agency) loans, suppliers' credit or other concessional loans taken from multilaterals especially for investing in the power sector.

In the field interviews conducted by the author it was revealed that the country is giving top priority to infrastructure development, including power where large capacity improvements have been achieved, largely from private sources, and this helping to draw foreign investment.

4.5.2 Gas Transmission Line

This is the second most critical factor in the economic category. Natural gas is the primary source of fuel for power generation in Bangladesh. Through Bangladesh is rich in natural gas, the present reserves are depleting quickly due to its wider use in the industry, most notably in the power sector. However, the present domestic gas reserve is expected to be completely exhausted by 2028–2040 unless new exploration and development work continues. Presently gas is not available all over the country and there is a shortage of transmission lines in the South, West and the Northern parts of the country.

In the field reviews it was found that the majority of foreign investors rated this factor quite high in considering the choice of fuel mix as part of their investment decision. Moreover, the price of natural gas is highly subsidized in the country, which is one sixteenth of the international level. This creates a proclivity to run power plants on this cheaper and cleaner fossil fuel which also produces fewer greenhouse gas (GHG) emissions. Evidence suggests that it is easier to get access to more sources of financing for such projects. Moreover, as the government has started to import LNG from 2018, which would be fed through the existing gas transmission pipelines, and more pipelines need to be built in the future. This would create more opportunities to invest in gas-based projects in Bangladesh [18].

4.5.3 Government Spending for Infrastructure

Bangladesh's recent economic development has been strong. From 2012 to 2019 its GDP grew at an average rate of 7% per annum. Bangladesh is on the track record to graduate from the United Nation's (UN's) least-developed countries in 2024. To meet its rapid development growth the government has increased its infrastructure investments since FY 2016. According to the government's 7th Five-Year Plan (2016–2020) infrastructure finance needs to reach 4.2% of GDP per annum or about US$ 14 billion. According to the revisiting PSMP 2016, it is estimated that to meet upcoming demand and maintain reserve margin, the country's generation capacity needs to increase by 74.5 gigawatts from 2020 to 2041. In line with this development the government has increased its ADP allocations for the power sector from Taka 24 billion in 2021 to Taka 502 billion in 2025 [35]. It could be suggested that ADP allocations in the power sector has been continuously increasing from the Sixth Five-Year Plan from 3.2% of GDP in 2010 through 5% of GDP to 8.7% of GDP in 2020. Out of the total ADP allocation in the power and energy sector in the Seventh Five-Year Plan, power sector itself received 53% of the overall financing for implementing the government's generation targets through a continuous progress in mobilizing IPPs and FDI in this sector.

In the field interviews it was revealed that the government's significant involvement in the power sector is a strong determinant for attracting private investment. However, at the project level some inconsistencies have been reported. For example, for many foreign investors the infrastructure for setting up power plants had to be fully supported by them with minimal involvement from the government. This is especially noticed in unsolicited offers that did not go through the usual tendering process. This included from purchase of land to developing the infrastructure during construction, installation and maintenance of power plants, especially installing jetties and fuel storage facilities to building transmission lines for power evacuation connecting to the load centers. This is more pronounced for all renewable companies awarded through uncompetitive solicitation.

4.5.4 Financial Facilities

This factor is rated high in attracting FDI in Bangladesh's power sector. In Bangladesh typical loan tenors from local commercial banks are in the range of 5 to 7 years and are in relatively small amounts. Moreover, banks require equity of 25%–35% of project costs for financing power sector projects. Club financing and syndication have been the norm with the largest syndication to date being $ 57 million in the power sector. Interest rates are also high at around 12.5%. Projects larger than US $ 70 million to US $ 100 million are difficult to finance domestically [36]. In addition, the corporate bond market and the capital market are not well developed and lack efficient risk transfer mechanisms for hedging in long-term power sector investments [37]. On that basis, international bank finance has been the norm for major source of project financing for developing power projects in Bangladesh.

It was revealed from the field observation that there is a combination of local and foreign financing for power projects including taking credits from multilateral agencies like the Asian Development Bank and International Finance Corporation (IFC), a lending arm of the World Bank for development projects. In addition, the government non-banking financial institution IDCOL had been instrumental in help arrange large credits for their respective FDI power companies, which ranged from US$ 15 - US$ 80 million for medium to large projects.

4.5.5   Credit Facilities

Foreign investors can utilize multiple sources of credits to finance their power projects in Bangladesh. These are financing from multilateral agencies, non-banking financial institutions, investment promotion and financing facility, local banks and foreign bank finance. The Bangladesh bank increased its single borrower exposure limit from 15% to 35% for funding infrastructure projects including power plants and have allowed foreign borrowing in the forms of financial loans, bank loans, buyer's credit, supplier's credit and debt issues from international financial markets [38]. Currently the central bank has allowed to extend credit to power sector borrowers exceeding the 25% limit as the cost of power generation has increased due to rise in dollar prices for import of fuel and raw materials from abroad [39].

However, the interest rates on such foreign borrowings are much lower than the local commercial banks. In addition, Bangladesh government in certain circumstances have allowed sovereign guarantees provided by partial risk guarantees by international financial institutions for enabling international financing of power projects. Though there is limited scope of large-scale financing by the local capital markets for power generation projects as the country lacks a developed capital market including a liquid corporate bond and a swap market. Moreover, due to limited scope for local commercial banks for arranging large credits for financing power projects, foreign investors have to rely on club financing and syndication to increase pooled finance but this too has limitations as the largest syndication to date to finance power projects had been US $ 57 million which is considered small given the large capital outlays involved in funding power projects [24]. Moreover, the loan tenor is short and the interest rate is quite high.

It was revealed that the locally available sources of finance are not adequate to fund power projects and foreign investors had to increasingly rely on foreign sources of finance for mobilizing large capital for their power projects. Though IDCOL has been instrumental for arranging large finance for power project but its interest rate is quite high from local commercial bank borrowing.

4.6   Political

4.6.1   Coordination and Collaboration Between Ministries

This factor is the most crucial factor in the political category. It was revealed that there is a lack of effective coordination and collaboration among the responsible agencies that constitute the power sector. This results in burdensome bureaucratic procedures and lack of transparency while providing business set-ups and other facilitation services to foreign investors. Especially there is not a single "one stop shop" that coordinates with the different line ministries, government agencies and departments for the speedy issuance of permits and licenses. This results in power companies to recruit in additional manpower or using dedicated resources to negotiate such services resulting in project delays and incurring additional cost.

4.6.2 Accountability of Public Officials

This assumes the second most important factor in the political category. This has significant bearings on the time, preparedness, and the costs of foreign investors. For example, in a 2008 bid involving two pre-selected large gas-based power plants (i.e., the Serajganj 450 MW and the Meghnaghat phase III 450 MW), the bidding process was cancelled several times. This was due to lack of a proper market assessment for the availability of gas, which led to a revised technical configuration based on dual fuel and a new tender. However, when the investors complied with the new scope and nature of work, their costs were considered too high, and the bid was cancelled. This kind of activity seriously impacts the costs and project preparation activities of the investors and dampens their enthusiasm for participating in the bidding process of large IPPs in Bangladesh.

4.6.3 Control of Corruption

Corruption is a major issue that prevents foreign investors from participating in Bangladesh's power sector. Irregularities and rent capture are considered significant stumbling blocks at the procurement stage. For example, procurement irregularities have resulted in persistent problems about procedures and transparency with the World Bank, especially with respect to large projects with multilateral involvement.

Additionally, projects are sometimes blocked at the final stage by rival factions or disaffected parties and are called for additional re-tendering. These factional conflicts are common and influence the award of private power projects in Bangladesh, causing significant rent capture and corruption at the procurement stage, favoring particular sponsors who represent powerful party elites [36].

Moreover, it was revealed that after getting contract awards, firms expend considerable time and resources to build informal personalized relationships with government officials in the form of paying rents or offering bribes. Such relationships help avoid any penalties and any supply shortage by having a good record on file.

4.6.4   Capacity to Adapt Policies

This factor is rated as very important to conduct FDI in the Bangladesh's power sector. It is believed that when the government has taken up a new policy due to global concerns or the local development of the sector such as a global response to reduce GHG emissions or a shift of natural gas to greater use of coal for the future capacity expansion of the sector, or the increase of renewables in the energy mix and adopting energy efficiency and conservation measures, the government should adapt to these new changes and help build the requisite capacities. For example, the government has already started implementing some of these plans such as targeting 10% use of renewables in the energy mix by 2021, adopting the use of highly efficient coal fired power plants using supercritical and ultra-supercritical technologies coupled with the development of the requisite infrastructure, enactment of laws and policies and the creation of dedicated agencies for their implementation and monitoring. However, it is suggested that many of the implementation works are slow, lack detail methods and approaches to evaluate risk and develop action plans and some are not in line with the global best practices like the adoption of Feed in Tariffs (FIT) for the greater expansion of renewables. However, it is assumed that this helps to build the confidence from the foreign investors about the future direction of this sector and plan investment opportunities along with these developments.

4.7   Social

4.7.1   Citizen Security and Accountability

For social factors, only citizen security and accountability was ranked as a very important factor in affecting investor decisions to conduct FDI. In our interviews, most respondents expressed their apprehension about the overall security (both life and goods) conditions in the country. This may be attributed to the fact that the country's law and order situation (for example, crime, theft, and other disorders) has seriously deteriorated. Additionally, there have been kidnappings and killings of foreign nationals. The most vivid representation of this manifestation is the very recent terrorist attack conducted in the diplomatic enclave of the capital, where nine Italian and seven Japanese personnel were killed [40]. This incident has garnered wide-scale global attention about the current law and order situation in the country, and influenced the government to maximize security efforts for combating global terrorism. However, the country has seen significant improvements in fighting terrorism and it sought regional and international cooperation and the local police and the government ministries (i.e., The Ministry of Religious Affairs and the National Committee on Militancy, Resistance and Prevention) are working with religious leaders and scholars, including the general public and civil society to curb terrorism. In that effect the country experienced some sporadic instances of terrorism acts in 2019 (i.e., directed to the police and political leaders) which has significantly improved since 2016. Bangladesh has 'zero tolerance' policy for terrorism and continues to support itself as a terrorist free country. It has taken a wide range of programs including cooperative activities through country support mechanism and international organizations like the Global community engagement and resilience fund (GCEF) to support local grassroots anti-terrorism efforts for risk to minorities. And in 2019 it hosted the inaugural national countering violent extremism (CVE) conference in concert with the US Embassy in Bangladesh, the United Nations (UN) and other partners aimed at producing a national CVE strategy [41].

4.8 Summary

This chapter summarizes the key factors that are important in making investment-decision to attract FDI in the Bangladeshi power sector. It has presented the factors that are deemed 'very important' to 'extremely important' that determine investment decision in the power sector. These factors are categorized into the four broad areas of investment prospects namely, regulatory, economic and financial, political and social subsumed under the role of institutions that determine FDI attractiveness in the Bangladeshi power sector. Evidence suggests that MNEs do not give equal weight to all the four categories of investment dimensions. The regulatory assumes the highest weight which is followed by economic and financial, political and social respectively. In the regulatory category commitment to contracts is the most important factor followed by land acquisition/rent/lease of land, tax exemption, presence of government guarantee and world-class security package respectively. For economic and finance, economic growth and development has the highest rating followed by gas transmission line and skilled labor. For political, the coordination and collaboration of ministries is the most important factor which is followed by accountability of public officials, the control of corruption and the government's capacity of adapt policies. For social, citizen security and accountability assume the highest rating.

Appendix

Part 1: Semi-Structured Interview Questions for Finding the Factors and the Inhibitors Generating Power Sector FDI

Research objective: This study aims to investigate the factors that are perceived to be important for help generating power sector FDI and also the barriers that impede their sustainable development.

Date: _____ Starting   time: _____ Finish   time: _____

Firm's   name: _____ Position: _____

Interview Questions:
1. What is the basic profile of your company? The topics include (firm's size, firm's ownership (wholly owned subsidiary, equity joint venture, minimum equity participation), the number of employees, business strategies and goal, competencies, key customers, type of investor, (namely, IPP developer, strategic investor or a combination), industry segment (in Bangladesh's context specifically the generation segment for private investment).
2. How long have you been working in this firm and how long have you been working in your current position?
3. How important do you think FDI in Bangladesh's power sector?
4. What are the strategic factors that influenced you to make investment decision in Bangladesh's power sector?
5. What are the potential barriers you think that are inhibiting sustainable FDI generation in Bangladesh's power sector?
6. Apart from the proposed factors, could you also identify some additional factors or disregard some irrelevant ones that have been put forward in the discussion?
7. What are your recommendations for sustainable FDI generation in Bangladesh's power sector particularly from the regulatory point of view?

Part II: Personal Information

Please choose (□) the answer that is most appropriate of your personnel background.

1. Gender

□ Male □ Female

2. Age

□ Below 30 years □ 30–44 years □ 45–59 years □ 60 years or above

3. Education

☐ Below Bachelor's degree ☐ Bachelor's degree ☐ Masters degree ☐ Ph.D degree

4. Period of employment with your current firm

☐ Less than 1 year ☐ 1–5 years ☐ 6–10 years ☐ Above 10

5. Your current position

☐ Officer ☐ Head/Supervisor ☐ Specialist/Senior Officer/Engineer ☐ Manager ☐ Director ☐ Executive ☐ Other (please specify)

6. Your responsibility

☐ Administration ☐ Operation/ Production planning ☐Research & Development ☐ Marketing /Sale ☐ Purchasing ☐ Logistics & Warehouse ☐ Human Resource Management ☐ Other (please specify) _____.

7. Period of working in your current position?

☐ Below 30 years ☐30–44 years ☐45–59 years ☐ 60 years or above.

8. Recommendations:

_____
_____
__ _____
_____
_____

Part III: Factors/Determinants Influencing FDI Decision Making in Bangladesh's Power Sector

Please indicate (☐) to what extent the factors influenced your firm to make investment decision in Bangladesh's power sector. Please choose the most appropiate level ranging from (1) Extermely important to (5) Not at all important.

No.	Factors	(1) Not at all important	(2) Slightly important	(3) Fairly important	(4) Very important	(5) Extremely important
1. Regulatory						
Investment process						
1.	Competitive selection process					
2.	Responsiveness of needs and time frame of investors					
3.	Presence of government guarantee					

No.	Factors	(1) Not at all important	(2) Slightly important	(3) Fairly important	(4) Very important	(5) Extremely important
4.	World class security package					
5.	Government's commitment to contracts					
6.	Power and Energy Fast supply Enhacement Act					
7.	No international benchmark for tariff setting ☑ Others					

Establishment

No.	Factors					
1.	Construction permit					
2.	Time & Efficiency of staff to complete the procedure					
3.	Liability insurance					
4.	Qualification of personal who supervises construction ☑ Others					

Revenue Risks/Controls

No.	Factors					
1.	Tax exemption					
2.	Termination of contracts without compensation to foreign stakeholders					
3.	Price cap regulation					
4.	Subsidy for consumers					

Regulatory risks & controls

No.	Factors					
1.	Competition policy					
2.	Wholly owned subsidiary /JVs					
3.	Protection of Foreign investors Act (1980)					
4.	Long approval process of IPPs					
5.	Protection of property rights					
6.	Profit repatriation controls					
7.	Environmental regulation					
8.	Need for internationally accepted environmental and social impact assessment (ESIA) for large projects ☑ Others					

Government & legislative processes

No.	Factors					
1.	Level of administrative competence					
2.	Continuity & consistency of rules & processes ☑ Others					

No.	Factors	(1) Not at all important	(2) Slightly important	(3) Fairly important	(4) Very important	(5) Extremely important
Labor market						
1.	Wage and other returns					
2.	Worker's insurance					
3.	Employment condition/ turnover					
4.	Conduct towards female workers					
5.	Trade union					
6.	Health, hygiene and safety of workers ☑ Others					
Foreign investment						
1.	Land acquisition					
2.	Property registration					
3.	Tax/rebate scheme					
4.	Exit policy (for large projects 5–7 years) for smaller projects after 02 years)					
5.	Can participate in more than one project for prequalification of investors and/or tenders					
6.	Not obliged to sell foreign share through public issues					
7.	Can buy shares locally/acquire local company					
8.	Foreign technicians not subject to personal income tax for up to 03 years					
9.	Avoidance of double taxation					
10.	When investing their retained earnings/dividends locally will be considered as new investment					
11.	Quick allocation of work permits ☑ Others					
International trade						
1.	Free trade across border					
2.	Free flow of raw materials ☑ Others					
Financial institutions						
1.	Approval of central bank for transferring capital ☑ Others					

No.	Factors	(1) Not at all important	(2) Slightly important	(3) Fairly important	(4) Very important	(5) Extremely important
Judicial structure						
1.	Fast track procedure for small claims					
2.	Commercial arbitration governed by a consolidated law or chapter of the applicable code of civil procedure (Bangladesh Arbitration Act 2001) ☑ Others					
2. Political						
Voice and accountability						
1.	Democracy					
2.	Vested groups					
3.	Accountability of public officals					
4.	Civil liberties					
5.	Electoral process					
6.	Transparency in government policy making					
7.	Freedom of press					
8.	Military in politics					
9.	Violance and terrorism					
Government effectiveness in implementing policies						
1.	Capacity to adapt policies					
2.	Policy consistency and forward planning					
3.	Political interferences					
4.	Excessive bureaucracy and red tape					
5.	Coordination/collaboration between ministries					
6.	Control of corruption					
3. Economic and financial						
Economic factors						
Growth & income						
1.	Economic growth and development					
2.	Good investment credit rating by Moody's					
Government side						
1.	Government spending for infrastucture					
2.	Government debts					

No.	Factors	(1) Not at all important	(2) Slightly important	(3) Fairly important	(4) Very important	(5) Extremely important
Labor						
1.	Labor costs					
2.	Human capital/skilled labor					
Infrastructure						
1.	Gas transmission line					
2.	Deep seaport					
3.	Domestic waterway					
4.	Railroad					
5.	Coal and LNG terminal					
6.	Bulk oil terminal					
Prices						
1.	Inflation					
2.	Real exchange rate					
Financial						
	Financial facilities					
	Credit facilites					
4. Social						
Inclusion						
	Male predominance					
Cohesion						
	Strengthening links between citizens and the government and promote more accountable government structures					
Resilience						
	Resistance to natural resource extraction (like coal, oil or gas)					
	Citizen security and accountability ☑ Others					

References

1. Hoskisson R, Eden L, Lau CM, Wright M (2000) Strategies in emerging economies. Acad Manag J 43:249–267
2. Doh JP, Rodriguez K, Uhlenbruck K, Collins J, Eden L (2003) Coping with corruption in foreign markets. Acad Mark Manag 17:114–121
3. Dunning JH (1998) Location and the multinational enterprise: a neglected factor. J Int Bus Stud 29:45–66
4. F Hatem (1997) International investment: Towards the year 2001 United Nations Geneva

5. Dunning JH (2006) Comments on dragon multinationals: new players in 21st century global-ization. 23:139–141
6. Cresswell JW (2007) Qualitative inquiry and research design: choosing among five approaches. Sage
7. Sekarn U (2000) Research methods for business: a skill-building approach, 3rd edn. Wiley, New York
8. Lamech R, Saeed K (2003) What international investors look for when investing in developing countries. World Bank, Washington, DC
9. Beasant-Jones JE (2006) Reforming power market in developing countries: what have we learned? World Bank, Washington, DC
10. Stern J, Holder S (1999) Regulatory governance: criteria for assessing the performance of regulatory systems: an application to infrastructure industries in the developing countries of Asia utility. Policy 8:33–50
11. Fraser JM (2005) Lessons from the independent private power experience in Pakistan. http://www.documents.worldbank.org/curated/en/729661468285358780/Lessons-from-the-independent-private-power-experience-in-Pakistan. Accessed 24 Mar 2019
12. Leferve T, Todoc J (2000) IPPs in APEC Economies: issues and trends. http://www.egcfe.ewg.apec.org/publications/proceedings/CoalFlow/ThaiSeminar_2000/T_vre_all.pdf. Accessed 4 Jan 2022
13. Woodhouse E (2006) The obsolescing bargain redux: foreign investment in the electric power sector in developing countries. Int Law Polit 38:121–219
14. Ramamurti R, Doh JP (2004) Rethinking foreign infrastructure investment in developing countries. J World Bus 39(2):151–167
15. Contractor FJ, Nuruzzaman N, Dangol R, Raghunath S (2021) How FDI inflows to emerg-ing markets are influenced by country regulatory factors: an exploration study. J Int Manage 27:100834
16. Hossain M (2015) Improving land administration and management in Bangladesh. https://cepa.org.mw/Library/natural-resources/Improving%20Land%20Administration%20and%20Management%20in%20Bangladesh.pdf. Accessed 1 Jan 2020
17. UNCTAD (2013) Investment policy review. http://unctad.org/en/PublicationsLibrary/diaep-cb2013d4_en.pdf. Accessed 8 Aug 2020
18. Mahbub T Jongwanich J. Determinants of foreign direct investment (FDI) in the power sector: a case study of Bangladesh. Energy Strat Rev 24:178–192
19. Kathuria S, Maluache MM (2016) Strengthening competitiveness in Bangladesh-thematic assessment. World Bank, Washington, DC
20. Daily Star (2023) Double taxation avoidance mechanism in Bangladesh. https://www.thedai-lystar.net/law-our-rights/news/double-taxation-avoidance-mechanism-bangladesh-1996229. Accessed 13 Jan 2023
21. Rahman AG (2010) An easy conversion. https://www.thedailystar.net/news-detail-139870. Accessed 10 Aug 2022
22. The Financial Express (2012) Sovereign guarantee to IPP projects 2020 (244)
23. World Bank USAID (1994) Submissions and evaluation of proposals for private power generation projects in developing countries. https://documents.worldbank.org/en/publica-tion/documents-reports/documentdetail/386871468764964315/submission-and-evaluation-of-proposals-for-private-power-generation-projects-in-developing-countries. Accessed 24 Aug 2020
24. World Bank (2010) Impact of the global financial crisis on investments in South Asia's electric power infrastructure: India Pakistan Bangladesh. The World Bank, Washington, DC
25. Rashidul H Jagaran C (2022) New patents bill passed. https://www.thedailystar.net/business/economy/news/new-patents-bill-passed-2997386. Accessed 20 Sept 2022
26. Bangladesh Bank (2018) Environmental and social management framework. Investment Facilitation Centre, Dhaka

27. Ahammed R, Harvey N (2004) EIA in Bangladesh: evaluation of environmental impact assessment procedures and practice in Bangladesh. Impact Assess Project Appraisal 22(1):63–78
28. Bangladesh Bank (2011) Environmental and social management framework for investment promotion and financing facility, Dhaka
29. World Bank (2019) Bangladesh development update: towards regulatory predictability. http://documents.worldbank.org/curated/en/269241554408636618/pdf/Bangladesh-Development-Update-Towards-Regulatory-Predictability.pdf. Accessed 17 Oct 2019
30. Doing Business 2020 (2020) Comparing business regulation in 190 economies. https://documents1.worldbank.org/curated/en/688761571934946384/pdf/Doing-Business-2020-Comparing-Business-Regulation-in-190-Economies.pdf. Accessed 8 Sept 2022
31. MOLE (Ministry of labour and employment) (2006) Bangladesh Labour Act. http://www.mole.gov.bd/site/view/legislative_information/Acts-%E2%80%8D&-Rules, 2015. Accessed 23 Mar 2015
32. World Bank (2012) Towards accelerated, inclusive and sustainable growth-opportunities and challenges. http://documents.worldbank.org/curated/en/280061468006660483/pdf/NonAsciiFileName0.pdf. Accessed 10 Oct 2021
33. Maniruzzaman AFM (2003) The new law of international commercial arbitration in Bangladesh: a comparative perspective. Am Rev Int Arbitr 14:139–203
34. CPD (2021) Bangladesh economy in FY 2020–21: interim review of macroeconomic performance. https://think-asia.org/bitstream/handle/11540/14088/Bangladesh-Economy-in-FY2020-21-Interim-Review-of-Macroeconomic-Performance.pdf?sequence=1. Accessed 8 Sept 2022
35. GOB (2020) 8th Five Year Plan July 2020 – June 2025: promoting prosperity and fostering inclusiveness. https://policy.asiapacificenergy.org/sites/default/files/Eighth%20Five%20Year%20Plan%20%28EN%29.pdf. Accessed 7 Sept 2022
36. Khan M, Riley T, Wescott C (2012) Public-private partnerships in Bangladesh's power sector: Risks and Opportunities. http://documents.worldbank.org/curated/en/550271468209052051/pdf/860780WP0Bangl00Box382147B00PUBLIC0.pdf. Accessed 7 Aug 2016
37. Bangladesh Bank (2013) Financing in infrastructure and energy sectors and issuance of bonds in Bangladesh: problems and prospects Working paper series: WP 1306. https://www.bb.org.bd/pub/research/workingpaper/wp1306.pdf. Accessed 20 Sept 2022
38. CPD (2015) Bangladesh economy in FY2014-15 Third interim review of macroeconomic performance. https://www.amazon.com/Bangladesh-Economy-FY2014-15-Macroeconomic-Performance/dp/9843392612. Accessed 7 Sept 2022
39. The Financial Express (2017) BB to bend rules for banks to invest beyond limit. https://thefinancialexpress.com.bd/trade/bb-to-bend-rules-for-banks-to-invest-beyond-limit-1505761870. Accessed 8 Sept 2022
40. Tharoor I (2026) American is among 20 dead in terrorist attack in Bangladesh. https://www.washingtonpost.com/news/worldviews/wp/2016/07/01/terror-attack-in-bangladeshs-capital-should-surprise-no-one/. Accessed 3 Sept 2018
41. U.S. Department of State (2019) Country report on terrorism 2019. https://www.state.gov/reports/country-reports-on-terrorism-2019/bangladesh/. Accessed 20 Sept 2022

Chapter 5
Barriers

Abstract The chapter identifies the key barriers subsumed under the four broad categories of investment prospects i.e., regulatory, economic and financial, political and social that deter FDI in the Bangladeshi power sector. This chapter is structure as follows. Section 5.1.- gives a detailed description of the most influential factors that deter investment decision to conduct FDI in the Bangladeshi power sector. The factors are drawn from the original study conducted to identify the key factors that encourage or discourage FDI in the Bangladeshi power sector as enumerated in the methodology in Chap. 4. A separate questionnaire was used to identify the key barriers drawn from the original study and the results were interpreted through in-depth one-to-one interviews and a survey among the four target groups including private company personnel, government officials, multilateral agencies, and academics. See the questionnaire in Appendix 2.

Keywords Barriers · Corruption · Power · Bangladesh · Gas

5.1 General Results

The results analyzed from the questionnaire show multinational enterprises do not provide the equal weight for the four areas of FDI deterrents. Figures 5.1 presents the results derived from the survey.

In terms of areas, political is the most influential followed by economic and financial factors, regulatory and social (Fig. 5.1). Within political corruption is the most import factor with a rating of 3.48 followed by political interferences with a rating 3.45 and lack of accountability of public officials 3.40 respectively. For economic and financial, inadequate gas transmission line is the most important factor with a rating 3.42. For social, citizen security and accountability scored the highest and finally for the regulatory dimension, the land acquisition processes assumes the highest weight.

© The Author(s), under exclusive license to Springer Nature Switzerland AG 2023
T. Mahbub, *Encouraging Foreign Direct Investment (FDI) in Bangladesh's Power Sector*, SpringerBriefs in Energy,
https://doi.org/10.1007/978-3-031-27990-4_5

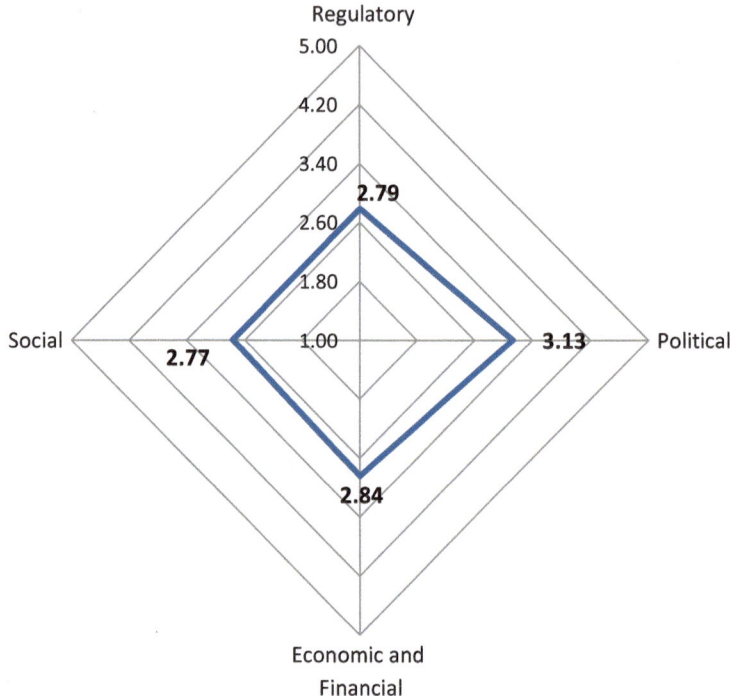

Fig. 5.1 Mean scores of influencing factors categorized by areas. (Author's estimates)

To facilitate the analysis, the means are interpreted as follows: (i) not at all important = 1.00–1.79, (ii) slightly important = 1.80–2.59, (iii) fairly important = 2.60–3.39, (iv) very important = 3.40–4.19, and (v) extremely important = 4.20–5.00. The following table 5.1 shows factors, which are classified as 'very important' ranging 3.40–4.19 which are reported here. From the four areas of investment prospects (i.e., regulatory, economic and financial, political and social) the following six factors are identified as 'very important' and are statistically significant at the 1 percent level.

1. Land acquisition/rent/lease
2. Corruption
3. Political interference
4. Inadequate gas transmission system
5. Long-approval process of IPPs and
6. Lack of accountability of public officials

Here the author would like to primarily highlight on three barriers which are (i) political interference; (ii) inadequate gas transmission system; and (iii) long-approval process of IPPs that are hindering the sustainability of FDI in the power

Table 5.1 Mean scores and P-values of individual factors

Class	Variable	Mean	P-value
Political	Corruption	3.48	0.000
	Political interference	3.45	0.000
	Lack of accountability of public officials	3.40	0.002
	Violence and terrorism	3.26	0.028
	Lack of coordination/collaboration between ministries	3.26	0.026
Economic and Financial (including infrastructure)	Inadequate gas transmission system	3.42	0.025
	Lack of coal and LNG terminal	3.27	0.040
	Lack of deep seaport	3.26	0.040
	Instability of labor costs	2.40	0.013
	Inadequate road network	1.60	0.000
Regulatory	Land acquisition/rent/lease of land	4.03	0.000
	Long approval process of IPPs	3.40	0.001
	Regulation on health, hygiene and safety of workers	3.27	0.007
	Delay in approval of central bank for transferring capital	3.27	0.022
	Lack of administrative competence	3.24	0.017
	Unresponsiveness of needs and timeframe of investors	3.19	0.020
	Foreign investors can participate in more than one project for prequalification of investors and/or tenders	2.47	0.041
	Termination of contract without compensation to stakeholders	2.47	0.042
	No international benchmark for tariff setting	2.42	0.010
	Foreign technicians not subject to personal income tax for up to 03 years	2.39	0.021
	Price cap regulation	2.35	0.002
	Regulation on subsidy for consumers	2.34	0.001
	Regulation on reinvestment earnings	2.27	0.001
	Regulation on acquisition	2.15	0.000
	Regulation on ownership (Wholly owned subsidiary/JV)	2.11	0.000
	Regulation on trade union	2.05	0.000
Social	Citizen security and accountability	3.26	0.000
	Resistance to natural resource extraction	3.16	0.026
	Male predominance	1.89	0.000

Source: Author's estimates

sector. It is to be mentioned that the other three barriers (i.e., land acquisition/rent/lease, corruption and accountability of public officials) have been highlighted in Chap. 4 as significant factors that influence investment decision-making to conduct FDI in the power sector.

Details are described below.

5.2   Principle Barriers Impeding FDI in the Power Sector

5.2.1   Long Approval Process of IPPs

This factor is highlighted as one of the key barriers in deterring FDI in the power sector. In Bangladesh power projects are subject to two kinds of bidding process. One is competitive bidding and the other is unsolicited or direct negotiation. In the open competitive bidding process, the off-taker normally uses a single-state, two-envelope bidding process where the bidder simultaneously submits two sealed envelopes, i.e., (i) containing the technical proposal and; (ii) the price proposal [1]. The price proposal will in the form of the bidder's respective guaranteed two-part tariffs for the sale of electricity from the power plants to the off-taker under the power purchase agreement (PPA). The tariff for supplying power is consisted of (i) capacity payment – this will cover debt service, return on equity, fixed operation and maintenance cost, insurance and other fixed costs; and (ii) energy payment which will cover the variable costs of operation and maintenance including fuel and is normally paid in Taka.

It is estimated that time required from the participation of a competitive bid offer till the award of the contract for simple power projects is 13 weeks and for complex projects[1] is 15 weeks, unless additional clarification is sought for technical, commercial and financial requirements, which adds to additional time and further delaying the award process for such bids.

The open competitive bidding process is well structured and conform to a list of criteria or checklists and specific guidelines that a project sponsor has to meet and the tender evaluation committee to follow for contract award. Its guidelines are enumerated in the request for proposal (RFP) for each solicited power projects conforming to open competitive bidding. Moreover, the open competitive bidding process is subject to the requirements of Public Procurement Rule (PPR) 2008 and the Public Procurement Act 2003, which specify the guidelines and requirements and timeliness for making contract award [2].

The second process is unsolicited bidding or direct negotiation. This falls under the purview of the Power and Energy Fast Supply Enhancement Act 2010 which has been further extended till 2021 for fast tracking power projects in order to meet emergency crisis situations for quickly procuring power and save the time on contracting power projects that would otherwise go through the open competitive bidding system. However, it was revealed that this process takes additional time, which on the average is 9 months for contact awards and there are several considerations for this, which are worth noting. One consideration is that the process itself is not

---

[1] This normally refers to two-stage tendering method covering procuring requirements for which it may not be in the best interests of the procuring entity to prepare complete technical specifications in advance due to rapidly changing technology or the procurement entity lacks the capability to prepare a full technical specification or alternative technical approaches may be available but not within the knowledge of the procuring entity.

structured as the investors do not know the likely outcome after submitting a proposal and the predictability of getting contract award. Another consideration is that there are no set specific checklists or guidelines that a project sponsor has to minimally conform to meet the technical, commercial and financial requirements of such projects in the initial proposal launching phase. This is subject to detailed clarifications and additional documentation requirements when the project when evaluating the proposal. A final consideration is that the process itself lacks transparency and is not subject to any time-bound screening requirements and evaluation criteria as opposed to open competitive bidding (where each phase is structured, time-bound and spelled out in detail) resulting in either fast awarding of contracts or long delays or risks of getting the proposal cancelled altogether. Moreover, it is argued that the sponsor has to intervene in all stages of the evaluation process in order to getting the responsiveness of the evaluators in each stage due to lack of timely intimation of outcomes or clarifications needed for the movement of the project through the successive stages before making a contract award. This further delays the award process.

5.2.2   Inadequate Gas Transmission System

Inadequate gas transmission system is considered a major barrier for FDI in the power sector. Bangladesh power sector is heavily reliant on natural gas, which constitute 71 percent of total generation followed by liquid fuel 13 percent, power import 9 percent, coal 4 percent and hydro 1 percent [3]. However, the domestic gas reserve is depleting fast and it has been estimated that gas production would peak by 2017 and would start gradually depleting thereafter. Presently, domestic supply of gas per year is 0.8 Trillion cubic feet (TCF), which would increase from 7–9 TCF by 2029. Presently, there is a shortage of 600 Million Square Cubic Feet Per Day (MMSCFD) as per demand, which is due to increased economic activity and new gas fields are not continuously explored both onshore and offshore for additional gas generation. Power sector and the industry including captive power consumes 69% of the gas, followed by transportation 8%, residence 20%, commercial 1.8% and agriculture 0.2% [4]. Presently gas is not available all over the country and there is a shortage of transmission lines with the majority concentrated in the central, northern and eastern parts of the country. The Government plans for expansion of the existing transmission lines with equitable and balanced development opportunities for the availability of gas in other parts of the country, which are falling behind with the gradual expansion of pipelines in the South, West and the greater Northern parts of the country [5]. Inadequate gas transmission line is a significant barrier for gas-based power companies. Especially in Bangladesh local gas is highly subsidized which is one sixteenth of the international market price draws the inclination for power companies to invest in gas-based projects for cheaper fuel. In addition, gas has less carbon emissions. Moreover, it is easy to get a wide source of financing for gas-based projects and easier to implement such projects for environmental considerations and cleaner power generation. Due to the present shortage of gas, the

Government plans to import LNG starting from 2017 and the existing transmission lines would then be used to feed LNG or develop new transmission networks for power generation, which is to be used in the existing gas-based government power plants or develop new gas-based power projects both public and private in the government's designated power hubs in the country.

5.2.3   Political Interferences

Political interference is considered as one of the principal barriers for sustainable FDI generation in the power sector. It is particularly manifested in the project procurement stage where powerful political actors have the capacity to influence decisions for project award or capture procurement rents from suppliers to effect contract decisions in their favor. In Bangladesh the Power Ministry is directly under the Prime Minister, which underscores the strategic importance of the sector. And the prime minister also works with the state minister and a power advisor. Anecdotal evidence suggests that powerful individuals both at the party and the bureaucracy also participate in the decision-making process. It has been argued that procurement rents can offer significant up-front rewards to those individuals and factions in the government and bureaucracy who are championing particular sponsors but this also comes at a price with high financing costs, adverse implications on the quality and future sustainability of projects and a higher risk premium over time [6]. A case of point to mention here is the award of the Bibiyana 341 MW gas fired project. The project was first offered through competitive solicitation to Summit Power Company. But due to the lack of arranging the requisite finance for the project in the specified deadline, the project was given to Marubeni-Hyundai, a Japanese-Korean joint venture through strong lobbying by a powerful lawmaker of the ruling party for which the tender authorities resorted to controversial means to eliminate the other qualified competitors. For example, the awarded company did not submit any authenticated or notarized end user certificate of experience, a compulsory requirement as part of prequalification, which normally disqualifies the company under a 'rejection clause'. Further, the off-taker sought 38 clarifications from the Marubeni-Hyaundai in order to qualify. Similarly, this also happened for the Haripur 360 MW power project where Electricity Generation Company of Bangladesh (EGCB) awarded the project to Marubeni based on using an unproven technology eliminating other qualified companies in the process. These illustrate the strong lobbying of political actors at work the award of IPPs in the frame of competitive offers [7].

5.2.4   Lack of Accountability of Public Officials

Accountability of public officials is generally seen as a significant barrier for FDI in the power sector. There are several instances of this, which are as follows. For example, contract awards are subjected to prolonged delays due to lack of timely decision-making and enforcement and the due delivery of commitments. For example, the bidding process of several large based-load power plants namely, Bibiyana, Meghnaghat Phase 2 and Sirajganj in 2009 was subjected to considerable delays, which culminated in a lack of sufficient response from the bidders, bid cancellations and further rebid of projects. This was due to lack of initial market preparedness, changing technical specifications at the early stage of the bidding process without prior intimation to the project sponsors, and the lack of a long-term fuel supply arrangement that dampened investors' confidence [8]. This has resulted in a lack of implementation of major new IPPs in Bangladesh since 2001. Moreover, there is significant rent capture at the project procurement stage, which creates factional conflicts both in the political apparatus and the bureaucracy resulting in protracted decision making and sometimes projects being blocked at the final stage by the disaffected parties.

In addition, there is frequent turnover of key management and staff in the sector which results in lack of ownership, interests and accountability in the decision-making process which bring significant delays in implementing power projects. A case in point is the funding of two World Bank funded investment projects in 2004 under the power sector development technical assistance project scheme that had serious implementation delays and gaps in governance which are worth noting. One consideration is that the projects experienced significant delays due to complications in implementation and interferences by several contracting agencies bypassing the Power Cell, the oversight agency under the power ministry to implement the projects. Another consideration is that project funds were channeled through two separate divisions of the power ministry instead of one agency requiring frequent meetings to review project processing and implementation which delayed the entire implementation process. Another consideration is that the corresponding power sector institutions involved with these projects tended to employ consultants to supervise the project contracts rather than engaging their own time and commitment to these tasks. Lastly, it took 4 years for the regulatory body; BERC, to access project assistance due to delays in the administrative apparatus of the Government. Especially due to frequent transfers and short time contracts of staffs in the sector making it difficult to retain staff with the right training, capacity and experience for implementing future power sector projects [9].

5.3 Summary

This chapter summarizes some of the key barriers that impede FDI in the Bangladeshi power sector. These barriers are (i) land acquisition/rent/lease of land; (ii) corruption (iii) political interference; (iv) long-approval process of IPPs; and (vi) lack of accountability of public officials. In terms of areas for FDI deterrence in the power sector the political dimension is the most important which is followed by economic and financial, social and regulatory. In the political dimension corruption assumes the highest weight. This is followed by political interference and lack of accountability of public officials. In the economic and financial dimension inadequate gas and transmission system assumes the highest weight. For social dimension citizen security and accountability is the most important and lastly, for the regulatory land acquisition/rent/lease of land assumes critical importance. A description of these barriers is given in this chapter based on on-ground examples garnered from the experiences of foreign investors from the field interviews when conducting FDI in the power sector.

Appendix

Part I. Obstacles/Barriers Influencing FDI Decision-Making in Bangladesh's Power Sector

Please indicate (□) to what extent these barriers negatively influenced your firm to invest in Bangladesh's power sector. Please choose the most appropiate level ranging from (1) Not at all important to (5) Extremely important.

No.	Factors	(1) Not at all important	(2) Slightly important	(3) Fairly important	(4) Very important	(5) Extremely important
1. Regulatory						
Investment process						
1.	Competitive selection process					
2.	Responsiveness of needs and time frame of investors					
3.	Presence of government guarantee					
4.	World-class security package					
5.	Government's commitment to contracts					
6.	Power and energy fast supply Enhacement act					
7.	No international benchmark for tariff setting ☑ Others					
Establishment						
1.	Construction permit					
2.	Time and Efficiency of staff to complete the procedure					
3.	Liability insurance					
4.	Qualification of personal who supervises construction ☑ Others					

No.	Factors	(1) Not at all important	(2) Slightly important	(3) Fairly important	(4) Very important	(5) Extremely important
Revenue Risks/Controls						
1.	Tax exemption					
2.	Termination of contracts without compensation to foreign stakeholders					
3.	Price cap regulation					
4.	Subsidy for consumers					
Regulatory risks and controls						
1.	Competition policy					
2.	Wholly owned subsidiary /JVs					
3.	Protection of Foreign investors Act (1980)					
4.	Long approval process of IPPs					
5.	Protection of property rights					
6.	Profit repatriation controls					
7.	Environmental regulation					
8.	Need for internationally accepted Environmental and social impact assessment (ESIA) for large projects ☑ Others					
Government and legislative processes						
1.	Level of administrative competence					
2.	Continuity and consistency of rules and processes ☑ Others					
Labor Market						
1.	Wage and other returns					
2.	Worker's insurance					
3.	Employment condition/ turnover					
4.	Conduct towards female workers					
5.	Trade Union					
6.	Health, hygiene and safety of workers ☑ Others					
Foreign investment						
1.	Land acquisition					
2.	Property registration					
3.	Tax/rebate scheme					
4.	Exit policy (for large projects 5–7 years) for smaller projects after 02 years)					

No.	Factors	(1) Not at all important	(2) Slightly important	(3) Fairly important	(4) Very important	(5) Extremely important
5.	Can participate in more than one project for prequalification of investors and/or tenders					
6.	Not obliged to sell foreign share through public issues					
7.	Can buy shares locally/acquire local company					
8.	Foreign technicians not subject to personal income tax for up to 03 years					
9.	Avoidance of double taxation					
10.	When investing their retained earnings/dividends locally will be considered as new investment					
11.	Quick allocation of work permits ☑ Others					
International trade						
1.	Free trade across border					
2.	Free flow of raw materials ☑ Others					
Financial institutions						
1.	Approval of central bank for transferring capital ☑ Others					
Judicial structure						
1.	Fast track procedure for small claims					
2.	Commercial arbitration governed by a consolidated law or chapter of the applicable code of civil procedure (Bangladesh Arbitration Act 2001) ☑ Others					
2. Political						
Voice and accountability						
1.	Democracy					
2.	Vested groups					
3.	Accountability of public officals					
4.	Civil liberties					
5.	Electoral process					
6.	Transparency in government policy making					

No.	Factors	(1) Not at all important	(2) Slightly important	(3) Fairly important	(4) Very important	(5) Extremely important
7.	Freedom of press					
8.	Military in politics					
9.	Violance & terrorism					
Government effectiveness in implementing policies						
1.	Capacity to adapt policies					
2.	Policy consistency and forward planning					
3.	Political interferences					
4.	Excessive bureaucracy and red tap					
5.	Coordination/collaboration between ministries					
6.	Corruption					
3. Economic and financial						
Economic factors						
Growth and income						
1.	Economic growth and development					
2.	Good investment credit rating by Moody's					
Government side						
1.	Government spending for infrastucture					
2.	Government debts					
Labor						
1.	Labor costs					
2.	Human capital/skilled labor					
Infrastructure						
1.	Gas transmission line					
2.	Deep seaport					
3.	Domestic waterway					
4.	Railroad					
5.	Coal and LNG terminal					
6.	Bulk oil terminal					
Prices						
1.	Inflation					
2.	Real exchange rate					
Financial						
	Financial facilities					
	Credit facilites					
4. Social						
Inclusion						
	Male predominance					

No.	Factors	(1) Not at all important	(2) Slightly important	(3) Fairly important	(4) Very important	(5) Extremely important
Cohesion						
	Strengthening links between citizens and the government and promote more accountable government structures					
Resilience						
	Resistance to natural resource extraction (like coal, oil or gas)					
	Citizen security and accountability ☑ Others					

References

1. ADB (2022) What bidding procedures are used by ADB-financed projects? https://www.adb.org/business/how-to/what-bidding-procedures-are-used-adb-financed-projects. Accessed 7 Sept 2022
2. CPTU (2008) The public procurement rule 2008. http://103.48.16.164/upload_images/content_images/Procurement%20Documents/03_Public-Procurement-Rules-2008-English.pdf. Accessed 8 Sept 2022
3. GOB (2020) 8th Five Year Plan July 2020 – June 2025: promoting prosperity and fostering inclusiveness. https://policy.asiapacificenergy.org/sites/default/files/Eighth%20Five%20Year%20Plan%20%28EN%29.pdf. Accessed 7 Sept 2022
4. JICA (2016) Survey on power system master plan (Draft final report). http://www.powercell.gov.bd/site/page/b93738b7-c6d2-4bcc-a7cd-298cbb91e6b4. Accessed 15 June 2016
5. Bhattacharjee P, Chong ZX (2021) Bangladesh central and southern regions will shape the future gas story, but pipeline access remains a bottleneck. https://www.ihsmarkitcom/research-analysis/bangladeshs-central-and-southern-regions-will-shape-the-futurehtml. Accessed 7 Sept 2022
6. Khan, M., Riley, T., Wescott, C. (2012) Public-private partnerships in Bangladesh's power sector: Risks and Opportunities. http://documents.worldbank.org/curated/en/550271468209052051/pdf/860780WP0Bangl00Box382147B00PUBLIC0.pdf. Accessed 7 Aug 2016
7. Khan S, Rahman MF (2011) Govt set to allow unproven techhttps://www.thedailystar.net/news-detail-191586. Accessed 3 Sept 2022
8. ADB (2011) Technical assistance completion report https://www.adb.org/sites/default/files/project-document/73710/41125-012-ban-tcr.pdf. Accessed 10 Sept 2022
9. World Bank (2014) Project performance assessment report. https://documents1.worldbank.org/curated/en/669251468212073331/pdf/885460PPAR0P070C0disclosed060240140.pdf. Accessed 6 Sept 2021

Chapter 6
Summary and Policy Implication

Abstract This chapter presents the conclusions and policy implications for sustainable FDI generation in the power sector both for conventional power sources (i.e., fossil-fuel) and renewable power generation for this sector. This chapter also presents areas for future research as a logical follow-up for the future direction of attracting FDI in the power sector.

Keywords GDP · Regulatory · Power cell · Fossil-fuel · Renewable

6.1 Summary

Bangladesh Government has targeted large expansion in power generation to meet the Vision 2021s target of having universal access to power by 2021 and becoming a middle-income country and beyond with the vision to becoming a developed economy in 2041. In that line it has projected a GDP growth on an average of 8% from FY 2021 to FY 2025 with significant budget allocations for the power sector ranging from Taka Billion 245 in 2021 to Taka Billion 543 in 2025 [1]. To meet the government's future growth targets of expanding capacity of 40 GW by 2030, it is estimated that a three-fold increase in generation capacity is needed continuing over the next 15 years. This would require US$ 40 to US$ 60 billion in capital expenditure over the next decade [2]. Since mobilizing these resources from the national budget is not sufficient, the government has targeted FDI in the form of Private-Public Partnerships (PPPs) and private foreign investment as a function to finance the large investment targets through the 7th Five-Year Plan and beyond to attain the objectives of the Vision 2021 for providing universal power for all and meeting the demands of its rising economy based on an accelerated export-led GDP growth.

Since the 1994, electricity sector reform has been implemented to break the state monopoly into separate generation, transmission and distribution systems and the promulgation of the 1996 private power generation policy with the requisite policy incentives to attract private and foreign investment. However, despite the government's increased promotional efforts and the adoption of attractive policy incentives, the actual growth of foreign investment in this sector has not been satisfactory as enumerated in the introductory chapter of this book. As a growing developing country, which has already attained the lower middle-income country status, Bangladesh has a rising need for power which is manifested in a growing demand supply and gap, which continues to increase with the pace of its rising economy. FDI is crucially important in the power sector, which is seen as a catalyst for attaining this rising GDP growth as research suggests that electricity use has a positive correlation to the level of economic development of a country [3].

Therefore, this study aimed to identify the factors that attract foreign investors to conduct FDI in the Bangladesh power sector and also identifies the barriers that inhibit investors to conduct FDI in this sector.

Evidence suggests that in the four areas of FDI attractiveness, i.e., regulatory, economic and financial, political and social foreign investors do not give equal weight to all these areas. The regulatory dimension is the most for FDI attractiveness which is followed by economic and financial, political and social. In terms of regulatory government government's commitment to contracts is the most important factor followed by land acquisition/rent/lease of land, tax exemption and avoidance of double taxation respectively. For economic and finance, economic growth and development has the highest rating followed by gas transmission line and skilled labor. For political, the coordination and collaboration of ministries is the most important factor followed by accountability of public officials and control of corruption respectively. For social, citizen security and accountability has the highest weight. In terms of identifying the factors that deter FDI generation in the Bangladeshi power sector in terms of areas, political is the most important which is followed by economic and financial, social and regulatory dimensions. In terms of political, corruption is the most important factor followed by political interferences and lack of accountability of public officials respectively. For economic and financial, inadequate gas transmission system is the most important. For social, citizen security and accountability assume the highest weight and for regulatory land acquisition/rent/lease of land assumes the highest score.

The empirical findings that are described here could be replicated in the South Asian and other developing country context for the identification of factors that are perceived to be important to conduct FDI in the power sector and the barriers that impede FDI in this sector and better inform governments on how to improve the conditions for sustainable FDI generation in their respective power systems. Especially those countries which have done functional unbundling of their respective power sectors into separate generation, transmission and distribution companies working under a single buyer model with the introduction of IPPs under long-term power purchase contracts for encouraging competition and are transitioning from a regulated market to a deregulated multi-buyer and multi-seller system aiming to develop a wholesale power pool. Another suggested alternative could be to having the provisions of a both regulated and competitive markets side by side with the arrangement of IPPs with take or pay contracts including merchant power plants participating in a bid-based spot market simultaneously if not completely thus transitioning to a competitive market arrangement if extensive unbundling and privatization of the country's power system is not politically desirable.

6.2 Policy Implication

The author would suggest the following policy implications for the sustainability of investment for attracting large-scale FDI in the Bangladesh's power sector. The following recommendations are proposed.

6.2.1 Making the Bid Evaluation Structured and Time-Constrained

The Private Power Generation Policy 1996 and later amended in 2004 has strictly mandated the Power Cell under the Ministry of Power and Mineral Resources to lead IPP projects. In that regard it will solicit and evaluate proposals, negotiate and process award of contracts and finalize various agreements. In addition, Power Cell will also facilitate all stages of promotion, development, implementation, commissioning and operations of IPPs and address any concerns of IPP sponsors. It is supposed to act as a one-stop agency fully dedicated for the processing of IPP proposals on an open competition basis based on a least-cost option-based power generation. Though it is meant to be the implementation agency for new private power generation projects and lead private power development, however, in practice, it is subject to frequent intervention by the beneficiary power utilities, which tend to bypass the Power Cell and implement their activities directly. This creates additional time and interference for project award and sometimes this is stretched beyond the bid validity phase for additional extension of time with considerable impact on time and costs on the IPP sponsors [4]. Moreover, this unit is subject to the Ministry's intervention and has not been allowed to function to its full form. For example, the BPDB the off-taker and the principle contracting agency for IPPs is supposed to facilitate the Power Cell for procurement and contracting IPPs on its behalf. The basic premise for having the Power Cell for soliciting new IPPs and their implementation is that for having the single buyer the BPDB to be a separate entity in the procurement process in order to prevent any potential self-dealing with IPPs for rent capture and keeping the public interest in the forefront to procure power using the least-cost investment options. Thus, having a check on any irregular activities or corruption which would prove to be a conflict of interest if an entity was responsible both for identifying least-cost investment options for public procurement and also benefitting through the process such as capturing monopoly rents from IPPs for having checks on such kinds of conflicting activities for the best interests of public procurement.

However, the Power Cell seriously lacks in manpower and the technical, commercial and the financial expertise to negotiate and implement projects on its own and has to depend on the off-taker the BPDB for having such auxiliary assistance. This is due to past and present governments have not allowed the agency to grow and allow the necessary resources and the empowerment to act as a single dedicated agency free from other interagency interference and vested group interests to implement private power generation projects. Moreover, apart from having a shortage of manpower and lacking in capacity, there are frequent transfer of staffs which cause serious impediments in retaining the training and experience to function as a single team learned through from project implementing experience to leverage future sound investment decisions in private power procurement [4].

From the interviews conducted with respondents, it was generally agreed that Power Cell's capacity should be improved by giving the agency adequate manpower, resources, training for developing the technical, commercial and financial expertise to review, monitor and negotiate proposals and help fully develop as an entity with performance targets for conducting new private power project implementation. The staff should have predictable tenure and the incentives and empowerment to act within the mandates of the Private Power Generation Policy without interference and represent the government's best interests in developing new private power generation projects.

On a counter note, some respondents also recommended that if the Power Cell is not allowed to fully develop itself to carry out its mandated functions as a single dedicated agency for IPPs, as it presently seriously lacking the institutional capacity to fully function on its own then the off taker should be allowed instead of the Power Cell to take over supervisory function of implementing IPPs. As it is fully developed with the right manpower, expertise and training to oversee new power generation procurement activities. In this regard, the existing policy should be modified so as to enable the concerned utility the BPDB to implement IPPs in the future. However, there should not be two parallel agencies namely the Power Cell and the BPDP for implementing IPPs, which is causing the present delay in IPP awards.

6.2.2   Use of Power Hubs

As access to suitable land is a major obstacle for foreign investors to conduct FDI in the power sector in Bangladesh owing to scarcity of land and the difficulty and complexity in the land acquisition process, foreign investors typically depend on the government for finding suitable lands for their ventures. Due to this perennial problem for acquiring land in suitable locations for the construction of power plants, as power plants consume large amounts of land (especially for renewable power for wine and solar energy projects) and the concomitant infrastructure (for example, fuel delivery infrastructure, proximity to water sources for cooling and makeup water, power evacuation system, proper system voltage support and the suitability of the ground and geotechnical conditions of the construction site), Bangladesh government has targeted power hubs for the best utilization of land for generating electricity. Presently there are a number of power hubs operating in the country which are: Ghorashal, Baghabari, Goalpara, Siddhirganj and Meghnaghat. The government has targeted to increase the capacity of these power hubs through the restructuring of the existing power plants, building new power plants both government and private and the establishment of necessary facilities without acquiring new land in the process. In addition, the government is planning to develop new power hubs under construction namely, Moheskhali, Matabari and Pyra [5] in the southern coastal belts of the country where the government has targeted to build several large coal-fired projects using ultra-supercritical technology and LNG projects for both government and private with a combined capacity of 10,000 MW as new power hubs come online in the future [6]. These power hubs would be structured in several blocks and will accommodate jetties, coal and LNG terminals and construction of townships for residential and non-residential facilities for each individual block along with a deep seaport for the transportation of coal into the power hubs [6]. As it has been estimated that 50% of the country's total electricity would come from coal-fired power plants by 2030.

In this regard the government is planning to finalize a power hub master plan in order to utilize the best use of land and develop the required infrastructure facilities for accommodating future power plants in the existing and the planned new power hubs sites. The use of power hubs will give foreign investors an integrated one-stop solution for having all the requisite infrastructure including land, access to fuel-supply infrastructure, transportation and delivery, water for cooling and the power evacuation system and the other needed facilities in place to set up power plants. Moreover, investors would get ready access to land in suitable strategic locations on a lease or rent basis by the government through having their power plants set up into these power hubs instead of acquiring land on their own, which proves to be a significant obstacle for foreign investors and discourages FDI.

For renewable power companies, evidence suggests that the Bangladesh Economic Zones Authority (BEZA) has been mandated to establish large number of private economic zones in the country for attracting FDI in solar and wind energy. These private economic zones will house solar and wind parks that would connect renewable power companies through an integrated infrastructure and transmission system by ready 'plug and play' for power generation. Additionally, owing to the scarcity of land a greater potential has been found to develop rooftop solar across industrial and commercial properties for which FDI companies are welcome in the future [7].

6.2.3 Strengthening the Central Procurement Technical Unit (CPTU)

The CPTU was established in 2002 under the Implementation, Monitoring and Evaluation Division of Ministry of Planning under the Government of Bangladesh as a unified national procurement body institutionalizing the procurement management capacity of the government and ensure transparency, fairness, efficiency and a better value for money in public procurement. It is mandated to follow the e-GP roadmap to fulfill the requirements of public procurement act 2006 and the public procurement rules 2008 in the solicitations of government procurement around the country under the different government agencies. With the establishment of the e-GP roadmap different procuring agencies can send their bid notices online on the CPTU e-GP platform and bidder can directly submit their proposal for the requested bid online and the whole cycle of procurement will be done electronically with the dissemination of information and monitoring done on a single electronic platform. Electronic procurement is regarded as an effective way to curb corruption in government procurement and ensure transparency and accountability. In its full functionality, it will support an e-contract management system, which will cover complete contract management processes like workplan submission, defining milestones, tracking and monitoring progress, generating reports, performing quality checks, generating running bills, vendor rating and generation of a complete certificate. Moreover, it will support to conduct real time dynamic pricing, which will enable prospective bidders to view opportunities, register expressions of interest, receive information and submit their proposals in a more informed manner.

In its initial phase the government targeted four agencies namely, Roads and Highways Department (RHD) Local Government Engineering Department (LGED), Rural Electrification Department (REB) and Bangladesh Water Development Board (BWDB) under Public Procurement Reform Project (PPRP-II) supported by the World Bank for electronic government procurement (e-GP) on a pilot basis that will help establish effective monitoring and evaluation for online platform standardization and carrying out procurement through online bidding document templates and processes. This will then help to carry out e -procurement for all government agencies under the different ministries on a regular basis. In addition, the government would launch a communication campaign for building social awareness on the benefits of e-procurement by engaging the civil society, outside independent organizations and other non-profit organizations in their role as a vigilance agency for effective monitoring and strengthening public procurement [8]. In that regard, the government has launched a training program involving the various stakeholders and would train about 11,000 people of the procurement community, which would involve policy makers, senior government officials, procurement officials, bidders and private individuals, field-level procurement staffs, civil society and the media for raising awareness of the new procurement system, the social benefits for the good value for money in public procurement, transparency, accountability and fair play in public procurement. And enhancing the capacity building of the CPTU to be fully operational in the coming years.

Apart from introducing e-GP, CPTU is also working on the strengthening of the PPR guidelines and mandates that all agency public procurements of different scope and size be strictly compliant with the PPR, which runs from calling tenders, use of standard bid documents across all public procurement, evaluation of the bids, timeline for completing the bid evaluation and project award, including auditing, handling of complaints and appeals, code of ethics and advertisement of awards [9].

In addition, it is supposed to incorporate the latest international practices with due amendment from time to time in the public procurement rule (PPR) and advice the different government agencies regarding the interpretation and implementation of these rules and inform on preparing tender documents and other assistance to procurement entities for improving governance in public procurement. Especially, shortening bid evaluation time and contract award, avoidance of rebidding and most important, penalize bidders for fraud in the procurement process including having them listed on the CPTU Website and the cause of debarment including informing the relevant authorities. Lastly, the agency is authorized to take actions against government officials who are engaged in corrupt practices in the procurement process in accordance to the government services rules and procurement regulations. It is suggested that the strengthening the CPTU in the future award of IPP contracts following the PPR and e-procurement guidelines would allow greater transparency and accountability in the procurement process and help curb corruption in public procurement in the future [9].

6.2.4 Having Highly Professional and Reputed People Doing the Tender Evaluation and Award

During the evaluation of tender award for IPPs at the procurement stage, the approving authority will help form an evaluation committee comprising of members from the procurement entity, in this case the off-taker, BPDB, the ministry and two outside experts from other reputed professional bodies or universities. Normally the bid evaluation committee would comprise a minimum of five members and not exceeding seven with two members outside the Ministry, the divisions or agencies under it. They should be experienced in technical, commercial, financial or legal matters for carrying out the tender evaluation and award [10].

It is recommended that the evaluation committee should be comprised of persons who are highly professional, dedicated and having high integrity and honesty when chosen for bid evaluation and award. Since the majority of the professionals in the bid evaluation committee are public servants of the relevant procuring entity or coming from the line ministry, therefore, from among them a cadre of procurement specialists should be developed who are highly motivated and having the right skills and integrity and who can effectively communicate and coordinate work with others for the best interests of public procurement. These people should be developed and deployed effectively for having the best value for the tasks entrusted on them in enhancing efficiency, competition, transparency and accountability in public procurement.

As integrity is the core element of professionalism, there should a change in the government's approach for procurement from a budgetary approach to a more managerial approach that provide more empowerment, discretion and flexibility to procurement officials to make the best-informed decisions regarding public procurement. In addition, their capacities should be improved by giving them adequate training in light of new regulatory developments, technological changes and providing a systematic approach to learning and development for help build and update their knowledge and skills.

The government should try to inculcate that public procurement as a strategic profession rather than a mere administrative function, which would instill a sense of pride and greater professionalism in the procurement officials and should reward them with adequate employment conditions and incentives for higher career prospects and personal development.

Lastly, to prevent the influence of vested groups on public decision making such as political elites who try to influence the procurement process for securing political rents and influence awards, procurement officials and other experts in the committee should strictly follow the public procurement guideline and its principles and should conform to the timelines set for each stage of procurement process namely, examination, evaluation and approval. As any extension of the timeline between each stage for seeking additional clarification from the parties sometimes set the stage for corruption for helping some preferred party to qualify and secure the award. Therefore, they should strictly abide by their terms of reference for professional conduct and ethics as enumerated in the public procurement rules for maintaining the due diligence and integrity and highest professional work ethics for conducting the tasks of project evaluation and contract award for the government.

6.2.5   Exploration of Onshore and Offshore Gas Fields

Natural gas is the primary source of fuel for power generation in Bangladesh. At present 71% of the total supply of gas is being used for generating electricity. It has been argued that the gas demand would peak by 2018 and start decreasing thereafter. According to Petrobangla, the national oil and gas exploration company the widening gap between demand and supply would be 7–9 Trillion Cubic Feet (TCF) by 2029 and the current reserve will be depleted in less than 10 years [11]. Out of the 27.12 TCF recoverable reserves, Bangladesh used 13.032 TCF natural gas by June 2015. And a further 14.088 TCF reserve remains for future consumption. Currently there is a shortage of 500 MMCFD as per demand due to low levels of exploration and in adequate transmission system. The current daily gas production by Petro Bangla is 1100 MMCFD and the rest is produced by the International oil companies (IOCs), which is 1600 MMCFD with a daily gas production of 2700 Million cubic feet per day (MMCFD). Evidence suggests that the government is planning to significantly increase the digging of gas wells to meet the demand of domestic energy consumption in light of the latest hike in fuel price due to the Russian-Ukrainian war. In this respect the government will add more 618 cubic feet of gas per day from 46 new wells that will be operational by 2025 [12].

It is recommended that Bangladesh should give top priority for future investment for exploration of new onshore and offshore gas fields. And subsequently LNG import should be considered to ensure smooth supply of natural gas. Since Bangladesh still has untapped gas resources and these are likely to lie in coastal-transitional areas or swampy-marshy areas, hill tract areas or offshore areas where future exploration is difficult and risky. It is suggested that Bangladesh should partner with International Oil Companies (IOCs) through attractive Production sharing contracts (PSCs) for conducting these kinds of costly and technically demanding explorations. Further, following the delineation of maritime boundaries between Bangladesh and Myanmar by International tribunal for the law of the sea (ITLOS) in 2012, a new opportunity has opened up for offshore for both shallow and deep-water explorations [6]. In this regard despite a weak response of the latest 2012 offshore bidding round, Bangladesh is preparing to offer a new round of bidding for offshore deep-water blocks with a more attractive PSC as competitive with those of neighboring countries like Myanmar and India to attract large scale international oil companies (IOCs) for future oil and gas exploration and development. Presently, IOCs are working only on a few shallow water blocks and not any deep-water blocks. It is suggested that even if gas was discovered it will take some time to have it commercially available as these are at a very early stage of development. Therefore, more emphasis should be placed on the exploration of the onshore fields at the present moment where already a significant reserve has been found. And in that regard the government has established a Gas Development Fund (GDF) where 15% of the gas tariff will be utilized for upstream exploration and development activities and currently the subsidiaries of Petrobangla are implementing projects with this fund and also financing through their own, especially for future onshore gas fields exploration and development.

Evidence suggests that currently the IOCs are reluctant to explore the offshore reserves as the PSCs are not attractive and the process of award or licensing of contracts are slow and cumbersome [13]. Since the conclusion of the 2008 offshore bidding round, the government has increased the contractual price of gas purchased from IOCs but the prices are seen as non-remunerative by the IOCs as the gas extracted from the field cannot be sold to third parties other than Petrobangla so that there is little interest in furthering more exploration in future onshore gas fields [1].

As Bangladesh has started to import LNG starting from 2018 and the cost of the LNG would be three times higher than Bangladesh's domestically produced and marketed gas, it is suggested that a price reform coupled with an attractive fiscal regime for domestic gas will be a key factor to attract private investors to explore and develop domestic gas fields. This would boost exploration and drilling of domestic gas, which would be cheaper than imported LNG in the future [1].

On the domestic front it is of critical importance to increase the domestic production of gas with greater exploration of the untapped onshore and offshore gas fields and the concomitant expansion of transmission lines all across the country. Especially some of the lagging southern, western and northern parts of the county where the gas supply is inadequate and lacking new transmission lines. This will help for a regional balance in energy supply all across the country as more gas becomes available with greater exploration and development works and also help implement new gas-based power plants in the future.

6.2.6   Development Plan and Renewable Target

In the Eighth Five-Year plan the government has laid considerable emphasis in increasing renewable power sources, i.e. wind and solar energy. As the cost of renewable technology is much cheaper due to a combination of maturing technologies, scale economies and competition and the growing consensus to have additional renewables in the energy mix including considering the social cost of carbon in investment decision-making are worth considering. The Bangladesh Government has set a target for renewable energy development including hydropower, solar, wind and biomass to 5% and 10% by 2015 and 2020. In addition, the government aspires to attain a target to achieve a 400 MW wind power capacity and another 100 mg utility-scale solar capacity as part of the government initiatives to tackle global climate change by 2030 as enumerated in its Nationally Determined Contribution (NDC) by 2030. Currently 716 MW of electricity are generated from renewable sources for which 543 MW renewable energy are under construction and another 1416 MW renewable projects are at the planning stage. It has been suggested that the government should continue to invest in renewable power generation and collecting and making available good data on the size and quality of solar and wind resources to support the development of the market including undertaking pilot and demonstration projects in renewable energy to attract confidence from foreign private investors to invest in renewables in the country. In this regard a sound renewable development strategy and targets with bottom up integrated energy planning that ensure consistency in upstream and downstream decisions and incorporating a coordinated sector planning that includes the different linkages (i.e., physical and financial constraints, availability of land, climate change, variabilities of weather and potentials for renewables) should help for developing a sound and resilient sector expansion path for the future [2].

In the optimum development of the renewable energy potentials and to attract greater FDI in this sector a transparent and publicly available renewable development plan needs to be in place that would give clear direction for both public and private stakeholders, conducting renewable energy resource mapping to confirm technical potential, identify suitable project locations or creating renewable energy development zones and infrastructure facilities for project locations and standardize renewable energy procurement.

6.2.7   Priority Access to the Grid

The current grid system in Bangladesh lacks the capacity and the sophisticated processes to accommodate large-scale renewables. This is an issue for both conventional fossil-fuel and renewable power companies (i.e., wind and solar) with the implications of grid failure to match the frequency of both sources in real time. Therefore, it is recommended that emphasis should be placed on smart grids with battery storage capacity for better integration of renewables. This would ensure a decentralized energy supply and bidirectional power flow for better system integration and uptake of variable renewables along with the conventional power supply. In addition, large-scale battery system should be emphasized for storing renewable energy for their ready dispatch at any hour of the day. This is important for renewable power companies as to the intermittency of renewable energies with variable sunshine and wind energy during the day for their optimum utilization and unhindered and continuing supply to the power grid. In this regard lithium-ion storage capacity has started to achieve economic price parity in many countries and could be a prospective investment choice for better integration of renewables with the grid [14].

6.3   Suggestions for Future Research

In ending, some suggestions are proposed for further research on the sustainable development of FDI in the Bangladesh's power sector. These are in the following.

- As Bangladesh plans to transit from a single buyer model to a multi buyer and multi seller system in the future as the market gradually matures with further unbundling, privatization and restructuring of the sector, future research could shed light on how FDI should be directed in this new structure and the nature of competition when such a system comes into place.
- As Non-OECD countries lead the growth in generation from renewables and by 2040, 63% of the power from renewables is generated in developing countries compared with 54% in 2014, future studies could also shed light on how Bangladesh plans to attract FDI in utility scale renewable power projects with state intervention and support mechanisms for greater deployment of renewables. And IPPs participation and pricing conducted through more transparent competitive open bidding processes toward attracting renewable FDI.
- As Bangladesh is in the move forward to introduce feed-in-tariff in renewables including policies for rooftop solar for scalable power development future research could shed light on how greater renewables FDI could be conducted in this sector based on generous support policies that are globally competitive for developing renewables.

References

1. GOB (2020) Eight Five Year Plan July 2020 – June 2025: promoting prosperity and fostering inclusiveness. https://policy.asiapacificenergy.org/sites/default/files/Eighth%20Five%20Year%20Plan%20%28EN%29.pdf. Accessed 7 Sept 2022
2. Pargal (2017) Lighting the way: achievements, opportunities, and challenges in Bangladesh's power sector https://esmap.org/node/174806. Accessed 4 Aug 2022

3. Ferguson R, Wilkinson W, Hill R (2000) Electricity use and economic development. Energy Policy 28:923–934
4. World Bank (2014) Project performance assessment report. https://documents1.worldbank.org/curated/en/669251468212073331/pdf/885460PPAR0P070C0disclosed060240140.pdf. Accessed 6 Sept 2021
5. Financial Express (2023) Three power hubs in the offing. https://thefinancialexpress.com.bd/national/three-power-hubs-in-the-offing-1557659810. Accessed 3 Aug 2022
6. GOB (2015) Seventh Five Year Plan FY 2016- FY 2020: accelerating growth, empowering citizens. https://www.unicef.org/bangladesh/sites/unicef.org.bangladesh/files/2018-10/7th_FYP_18_02_2016.pdf. Accessed 5 Oct 2022
7. Dhaka Tribune (2022) BEZA to set up 1000 MW solar electricity zone in Chandpur. https://archive.dhakatribune.com/bangladesh/power-energy/2017/06/23/1000-mw-solar-electricity-zone-setup-beza. Accessed 9 Sept 2022
8. World Bank (2022) World Bank-funded project drives public procurement reform in Bangladesh. https://www.worldbank.org/en/results/2022/08/04/world-bank-funded-project-drives-public-procurement-reform-in-bangladesh. Accessed 10 July 2022
9. Bangladesh Gazette (2011) Bangladesh e-Government (e-GP) guidelines. https://www.eprocure.gov.bd/help/guidelines/eGP_Guidelines.pdf. Accessed 10 July 2022
10. GOB (2008) The public procurement rules 2008. http://103.48.16.164/upload_images/content_images/Procurement%20Documents/03_Public-Procurement-Rules-2008-English.pdf. Accessed 15 Aug 2022
11. Petrobangla (2015) Annual report. https://www.resourcedata.org/dataset/rgi-petrobangla-annual-report-2015/resource/a16bedb3-3613-4632-bd3c-133f86d0da2a. Accessed 13 Aug 2022
12. Shawon AS (2022) Govt looking to ramp up domestic gas production with new wells https://www.dhakatribune.com/Bangladesh/2022/08/26/govt-looking-to-ramp-up-domestic-gas-production-with-new-wells. Accessed 10 Jan 2023
13. The Financial Express (2022) Importance of hydrocarbon explorations in offshore blocks. https://thefinancialexpress.com.bd/editorial/importance-of-hydrocarbon-exploration-in-offshore-blocks-1657983125. Accessed 10 Sept 2022
14. USAID (2021) Recommendation for a renewable energy implementation action plan for Bangladesh https://pdf.usaid.gov/pdf_docs/PA00XD5J.pdf. Accessed 20 Aug 2022

Index